Introduction to Mechanical Design

Hanyang University
Gyung-Jin Park

공리적 설계방법

東明社

공리적 설계방법(Introduction to Mechanical Design)

초판 발행 2012년 11월 30일
4쇄 발행 2017년 8월 15일

지은이 박경진
펴낸이 최국주
펴낸곳 동명사
출판등록 1950년 11월 1일(제1-76호)
주소 (우 10881) 경기도 파주시 회동길 50(문발동)
전화 031-955-7200(대표)~2, 7203(편집부), 7204(물류)
팩스 031-955-7205
전자우편 webmaster@dmsbook.com
홈페이지 www.dmsbook.com

ISBN 978-89-411-8111-8 93550
값 9,000원

Preface

Design education at the entry level tends to be quite important these days. Yet there seems to be a lack of general education courses. Moreover, it is difficult to find an appropriate textbook for such a course. Hanyang University offers an elementary design course. This volume is developed for the first half semester of the design course of Hanyang University. Therefore, it is not for professionals such as scientists and engineers but for beginners who have not yet been exposed to design. Professionals may want to read it to review elementary aspects.

A product is planned in the design process. Design can be viewed from two facets. One is the aesthetic viewpoint and the other is the scientific viewpoint. The beauty of the outer shape is emphasized in the aesthetic design while the logical process should be presented in the scientific design. Engineering designers generally use the latter method. Therefore, engineers need design methodologies that can reasonably explain the design process in an objective way. However, design methodologies are generally domain (discipline or problem) dependent or they do not have definite forms. Axiomatic design is selected as a representative method because it is domain free and has a simple process which is easy to understand.

This volume is not for a general explanation of design. It is specifically written for the elementary design course of Hanyang University. Chapter 1 describes what design is. Elements in the design process are introduced and explained. Chapter 2 briefly explains the elements in design and Chapter 3 introduces axiomatic design. Two axioms are presented and a few examples are demonstrated. In the second half of the semester, students conduct a design project by forming teams. How to write project reports are introduced in Chapter 4.

발췌본머리말

extract preface

Springer 출판사에서 나온 "Analytic Methods for Design Practice"란 책이 "설계방법론"이란 제목 번역되었다. 이 책은 "설계방법론" 중 공리적설계 부분만을 발췌한 것이다. "설계방법론"에는 많은 설계방법이 소개되어 있다. 그러나 대부분 어려운 이론을 요하므로 입문설계에서 사용하기에는 다소 어렵다. 입문설계에서 설계방법론으로서 공리적설계를 채택할 경우 이 책을 사용할 수 있으리라 생각한다. 몇 가지 이론 및 예제는 설계 입문자들에게는 어려울 것이다. 특히 정보량의 계산이나 그 의미가 까다롭다. 이 경우 어려운 부분은 지나칠 수 있으리라 생각된다.

박 경 진

머리말

preface

지난 20년 가까이 기계공학과에서 설계를 가르쳐 오면서 공부하고 배우고 느낀 점들을 요약하였다. 특히 지난 18년간 한양대학교에서 가르친 설계에 관한 사항을 정리하였다. 이 책은 책임저자가 Analytic Methods for Design Practice라는 제목으로 Springer 출판사를 통해 출간한 책의 번역판임을 밝힌다. 여기서의 설계는 주로 공학에서 많이 다루는 역학 계산의 결과를 제품의 계획에 반영하는 과정이다. 일반적으로 대상 시스템의 역학적 해석에는 많은 노력을 기울이고 있으나 이를 설계에 반영하는 과정은 즉흥적이거나 경험과 직관에 의존하고 있는 실정이다. 이러한 과정은 매우 비효율적으로 비추어질 수 있으나 실제로는 설계의 특징을 잘 나타내고 있다. 즉 설계과정은 오히려 역학적 계산보다 더 복잡해서 설계자의 사고나 판단으로 하는 것이 더 정확할 수 있다. 그러나 이 때문에 최선의 설계 결정은 객관적으로 증명할 수 없는 점도 있다.

최근 대상 시스템이 점차 대형화되는 추세가 있다. 대상 시스템이 크면 설계자의 주관적 판단만으로는 설계가 어렵고 객관적으로 검증할 수 있는 정형화된 설계방법이 요구된다. 이에 많은 설계방법을 개발하고 있다. 설계에는 필히 결정과정이 따르게 되어 설계자는 여러 개의 선택을 놓고 고민하게 되거나 하나의 결정도 못할 수 있다. 이러한 경우 설계자는 설계방법론을 필요로 한다. 즉 결정과정을 도와주는 도구가 필요하다.

설계는 대체로 제품의 전반적 특성을 정의하는 개념설계, 부품의 선정이나 배치 등을 결정하는 초기설계와 선정된 부품의 최종 치수를 결정하는 상세설계로 분류한다. 설계의 순서는 일반적으로 개념설계, 초기설계, 상세설계 순으로 진행한다. 이러한 모든 과정에 보편적으로 적용할 수 있는 설계방법이 있을까? 아직은 그러한 방법은 없는 것으로 보인다. 다만 설계과정에 따라 적용하는 방법을 달리 한다. 설계는 설계방법론의 적용 관점에서 전반적인 사항을 결정하는 개념설계와 치수를 최종적으로 결정하는 상세설계로 분류할 수 있다.

각 설계단계에 적용할 수 있는 설계방법들을 취급한다. 다시 말해 정형화된 설계방법을 설명하며 그들의 응용에 대해 논의한다. 이렇게 이론적 방법론이 확립된 정형화된 설계를 여기서는 해석적 설계방법이라 지칭한다. 각 단원은 하나의 방법을 다룬다. 우선 방법을 유도하고 설명한다. 예제를 통해 독자는 소개하는 방법을 연습할 수 있다. 실제적인 응용 사례는 단원 끝의 부록에 소개하였다. 따라서 각 단원은 독립적으로 사용할 수 있다.

이 책은 학부의 고학년이나 대학원과정의 수업에서 교과서로 사용할 수 있다. 배우는 학생은 기본적인 미적분에 대한 수학이나 공학의 기초 지식을 알고 있어야 한다. 또한 현장의 기술자들이 제품에 설계기술을 적용하려 할 때 사용할 수 있다. 설계방법만을 사용하려 한다면 이론적인 부분을 제외하고 설계방법의 과정만을 익히는 것으로 충분할 것이다. 최근 공과대학에서는 설계교육에 대한 요구가 증진되어 입문 설계, 캡스톤 설계, 창의 설계 등에 대한 논의가 활발하다. 이때 설계한 제품에 대한 객관적 검증에 여기에 소개되어 있는 설계이론을 응용할 수 있다.

각 단원을 간단히 소개하면 다음과 같다. 제 1 장에서는 설계의 전반적인 사항과 함께 이 책에서 소개하는 해석적방법들을 간단히 설명한다. 제 2 장에서는 개념설계에 유용한 공리적설계를 소개한다. 제 3 장부터 제 7 장까지는 상세설계에 적용할 수 있는 해석적방법들을 설명한다. 다루는 방법들은 최적설계, 구조최적설계, 동적반응최적설계, 실험계획법에 의한 설계, 강건설계이다. 제 8 장과 제 9 장에는 실제 설계에 방법론을 적용한 응용 사례를 소개한다. 끝으로 제 10 장에는 해석과 해석적설계법을 복합적으로 적용하는 다분야통합최적설계가 있다. 독자는 필요에 따라 각 장을 따로 공부할 수 있다.

한양대학교의 대학원에서 많은 박사와 석사를 배출하면서 해결한 예제들을 수록하였다. 고병갑 교수, 박영선 박사, 이권희 교수, 도성희 박사, 임재문 박사, 이광원 박사, 최우석 박사, 신문균 박사, 황광현 박사, 강병수 박사, 송세일 박사, 심진욱 박사, 박철민 박사, 이정욱 박사, 이재완 박사, 박기종 박사, 신정규 박사, 신광섭 박사, 전상기 박사들께 감사드린다. 그들이 없었다면 다양한 설계론을 접할 기회도 없었을 것이다. 특히 공동저자인 강병수 박사와 교정 등의 마무리 작업을 주도적으로 도와준 신정규 박사에게 감사드린다. 그간 배출한 70여 명의 석사과정의 제자들께도 감사드린다. 또한 현재 재학하고 있는 대학원생들께도 감사드린다. 책을 만드는 시기에 저자와 함께 있음으로 인해 교정 등에 각종 도움을 받았음을 밝힌다.

저자에게 석사과정에서 처음으로 설계를 가르쳐 주신 KAIST의 곽병만 교수님, 박사과정에서 심도 있는 연구를 가르쳐 주신 The University of Iowa의 Arora 교수님, 임관 교수님께 감사드린다. 부교수 시절 공리적설계를 가르쳐 주시고 연구하는 교수의 방향을 제시하여 주신 MIT의 서남표 교수님께 무한한 감사를 드린다. 무엇보다도 저자의 장인이신 The University of Iowa의 박준부 교수님께 감사드린다. 박 교수님은 부족한 사위에게 교수생활의 각 단계마다 무엇을 해야 할 지를 사랑으로 이끌어 주셨다. 현재도 많은 대화를 통해 조언을 아끼시지 않는데 앞으로도 바랄 것이다.

끝으로 사랑하는 가족들에게 고마움을 전하고 싶다. 우선 믿음으로 키워주신 부모님께 감사드린다. 특히 이 책의 최종 출간을 앞두고 돌아가신 아버님과 이제 연로하셔서 투병중이신 어머님께 자식의 이 자그마한 결실이 미약하나마 그간의 사랑에 보답이 되길 바란다. 미국에서 자라 아무 것도 모르고 저자를 따라 우리나라에 온 아내에게는 항상 빚진 마음이다. 그간의 내조에 혹시 작은 위로라도 되기를 바랄 뿐이다. 아울러 우리의 예쁜 상은이, 상화, 상우가 나중에 아빠의 책으로 공부를 할 수 있다면 더 바랄 것이 없겠다.

2007. 7.

안산 연구실에서

책임저자 박경진

Contents

차 례

content

제1장 설계기법의 개요

제2장 공리적설계

I. Introduction to Design

1.1 What is Design?

Design can be defined in a variety of ways depending on the specific context and/or the field of interest (Suh 2001). It can be an engineering activity, a plan for a product or fashion, a plan in everyday ordinary life, and so on. Therefore, the word "design" is utilized in a broad sense. In this volume, the meaning of design is narrowed to an engineering activity in the planning stage of a product. Since design is a plan, it is performed in the early stage of the engineering process.

The overall engineering process is illustrated in Figure 1.1. The total engineering process can be considered as design. In a narrow sense, the design stage resides after the overall planning and before the product is physically made. The corporate strategies are established and market objectives are defined in the

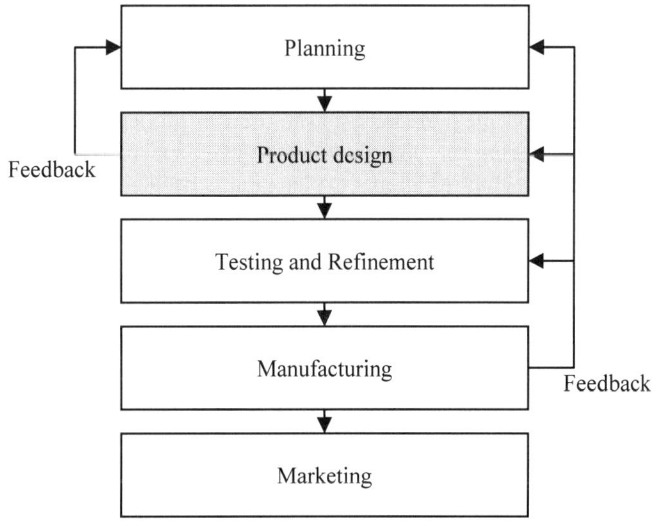

Figure 1.1. Overall engineering process

planning stage. The design activity has a bridge function between planning and manufacturing. In other words, a product is conceptually defined based on the problem definition of the planning stage. Although it occupies a small portion in the entire process, the impact of the design results is fairly large.

The history of design is extremely long. It has been performed since cavemen made tools for hunting. Although the history is long, a rigorous method with a definite theory is employed recently. Instead of a standardized method, intuition and unexplained experience have been utilized in design. Even now, there is some doubt as to whether or not there are any definite forms for design theories. Science based design methods are beginning to emerge.

Engineering design problems are usually open-ended and ill-structured (Dym and Little 2004). They are open-ended because there are multiple acceptable solutions and a solution can be improved upon. Uniqueness, which is extremely important in mathematics and many analysis problems, does not apply to design problems. In science and analysis, we generally use mathematical formulae in a structured way and obtain a solution. However, there are no such formulae in design. Due to the characteristics, creativity is involved in design because a different designer can make a different design. Engineering design shows the typical characteristics of engineering, which are different from those of sciences. It is noted that the aspects of design are utilized even in the scientific activity, especially when a plan for a scientific process is established.

A decision making process is typically involved in design. How do we make a decision or why do we make some specific decisions? What helps us make the decision? What would be rational and logical decisions? Who is a good designer and who is not? According to the manner of the decision making process, there are two viewpoints for the definition of design. In some designs, aesthetic beauty is emphasized. An example would be industrial design. In industrial design, visible shape or user convenience is important. Design is mostly made based on intuition; therefore, science based methods may not be used in this case. In other designs, soundness in engineering is considered. Achievement of target engineering functions is important. Thus, a rational explanation is needed. In this case, science based design methods can be used although it is not always the case.

This volume is prepared for college freshmen or sophomores who are not exposed to a specific discipline of engineering or science. They mostly learned general mathematics and sciences through high school education. It is assumed that the readers are familiar with analysis that uses a rigorous method and has a unique solution. The value system of design may confuse them because it is different from what they learned so far. However, it is quite important in that it is a real-world problem. Difficult terminologies and concepts are minimized; therefore, many aspects of this volume might be natural to professionals.

1.2 Analysis vs. Synthesis

The spectrum of analysis and synthesis is illustrated in Figure 1.2. Purc mathematics resides at the end of analysis while art is at the extreme of synthesis. Analysis pursues a unique solution of a problem in an analytical way while various aspects of sciences, intuition and artistic talents are synthesized to make an artifact in synthesis.

One of the characteristics in analysis is that there are solutions, in most cases a unique solution. The mathematics world is artificially defined with no exceptions; therefore, it is perfectly defined. In most cases, we can get a unique solution for a mathematical problem. Sciences try to formalize the phenomena of nature. Based on various assumptions, sciences employ the rules and laws of mathematics in the formalizing process and a unique solution is pursued. Scientific rules are exploited in engineering to analyze an artifact and a unique solution is still pursed in engineering analysis. The results of engineering analysis are reflected in design synthesis. Engineering design methods, intuition, creativity and experiences are utilized in this process. There can be various ways of reflection. Beauty is required in art. Human sense is utilized in art and almost infinite solutions may exist.

In analysis, the problem is often well defined and various mathematical as well as scientific principles are utilized. An objective approach is employed. On the other hand, intuitions, experiences and artistic sense are generally adopted in synthesis. A subjectivc approach is employed. Therefore, creativity is absolutely important in synthesis. Engineering design resides in between mathematics and art. Some design methods such as optimization heavily use mathematics and sciences. They are close to analysis and are not considered as design here. In some designs, intuitions and experiences are exploited. They are close to synthesis and generally considered as design. These days, there is a tendency to use an objective approach even in the design field which is close to synthesis. This approach is called a science based method.

A science based method should have principles. If we use such principles in design, we can obtain a design that is objectively good. Some designers doubt whether or not such principles exist. However, many efforts are undertaken to find

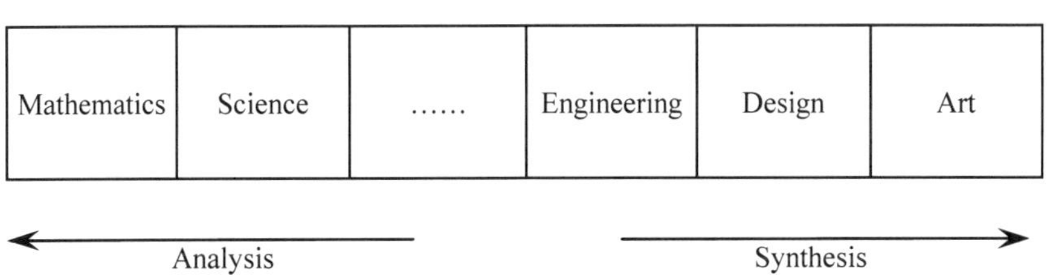

Figure 1.2. Analysis vs. Synthesis

good principles. One of them is the axiomatic approach and this will be explained later in detail. Anyhow we need design principles to reduce the lead time and cost as well as to assure the resultant design is good. Generally, the science based design methods are not as rigorous as mathematics. A unique solution is not guaranteed; however, there is a room for creativity because of that.

1.3 Design Process

Engineering design is the systematic, intelligent generation and evaluation of specifications for artifacts whose form and function achieve stated objectives and satisfy specified constraints (Dym and Little 2004). The role of design in the entire engineering process has been explained. In this section, the design process is investigated in detail. Design is classified in various ways depending on the viewpoints. It can be classified based on the order of the activity, design objectives, methods utilized, and so on.

According to the order of the activity, the design process is generally divided into conceptual design, preliminary design (embodiment design) and detailed design. In conceptual design, the problem is defined, data are gathered and overall parts and shapes are determined. A number of possible solutions can be defined and narrowed down to a single solution. Parts are selected and the relationships between parts are identified in preliminary design and detailed dimensions of parts are determined in detailed design. The order of these three designs is illustrated in Figure 1.3 and these are included in the product design stage of Figure 1.2. Since the direction of design is mostly determined in conceptual design, the impact of conceptual design is generally the largest.

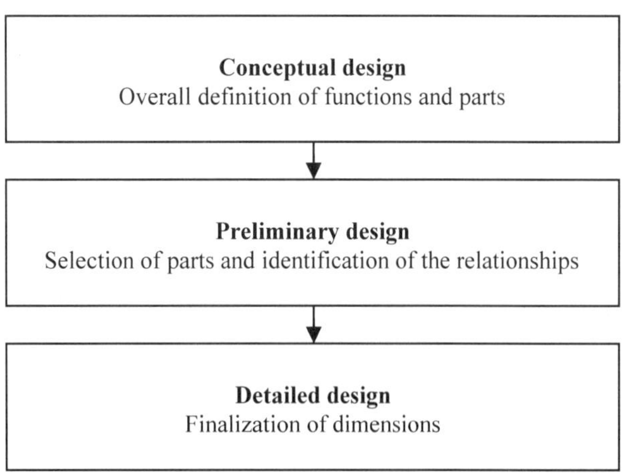

Figure 1.3. Sequence of design

Design can be classified into creative design and design improvement. A new product is defined in creative design. In the modern engineering community, new products are continually introduced in the market. Especially, many new electronic devices are made by information technology (IT). Creativity is important and intuition and experiences are exploited. Recently, designers tend to use design methods for creative design. Existing designs are usually improved in conventional technology. In this case, a new technology might be employed; however, the basic function of a new product is basically similar to that of the existing product. Generally, design methods with definite forms are being used.

Based on the design methods utilized, design is classified into aesthetic design, logical design and analytic design. Beauty and user convenience are pursued in aesthetic design. It seems that unexplainable intuition and experiences are the major sources in this case. The decisions should be rationally explained in logical design. Logical design is generally carried out in the conceptual design or the preliminary design of an engineering product. Mathematical processes are utilized in analytic design. It is exploited in the detailed design process. Analytic design is sometimes perfect in logic, therefore, it is close to analysis not synthesis. Although the logical process in decision making is excellent, the impact of detailed design is not large.

The entire design process consists of four domains such as the customer domain, the functional domain, the physical domain and the process domain. Figure 1.4 illustrates the order of the domains in the design process. First,

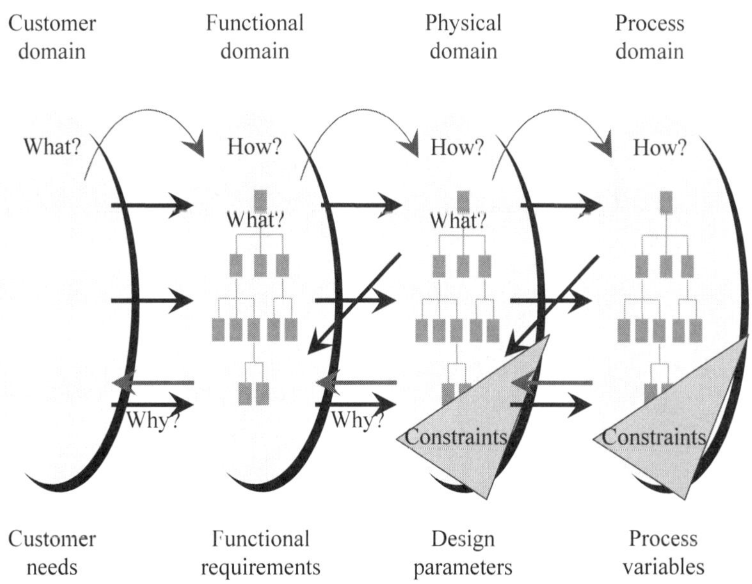

Figure 1.4. Relationship between domains, mapping and spaces

customer needs should be identified. Customer needs (CNs) are expressed by plain or abstract sentences. Engineers transform the customer needs to functional requirements (FRs). The FRs are expressed by engineering terminologies. This transformation is also called mapping. The FRs in the functional domain are materialized by design parameters (DPs) in the physical domain. Generally, the mapping process between functional and physical domains is called design. The designed product in the physical domain is manufactured in the process domain. Each step of Figure 1.4 uses the mapping process. The subsequent chapters explain the aspects of the four domains.

1.4 Exercises

1.1 Explain two viewpoints of design and show product examples of the two viewpoints. Do not use any references and use your own thoughts.

1.2 Describe analysis and synthesis with your own thoughts.

1.3 Show a design example that pursues aesthetic beauty and explain why.

1.4 Show a design example that pursues analytic objectives and explain why.

1.5 Explain what conceptual design, preliminary design and detailed design are. You can use references.

1.6 Explain the four domains of design with your own thoughts.

II. Elements in the Design Process

In Figure 1.4, the design process is demonstrated by four domains. The characteristics of the four domains are explained in this chapter.

2.1 Customer Needs and Market Trends

Customer needs (CNs) are demands or requests from customers. The customers can be general consumers when the product is for the public, and the customers can be an organization when a product is supplied to an organization. CNs from general consumers are quite abstract while those from an organization are specific.

In many cases, general customers do not exactly propose what they want. They need some good products; however, they do not express exactly what those are. Thus, engineers should usually identify CNs. CNs can be defined by a customer survey. The survey should include items that the customers want since customers usually do not think of anything else other than the items in the survey. There are various methods of how to make the survey and how to evaluate the survey results (Dieter 2000).

These days, designers often create customer needs. In other words, a designer can create a product which general customers could never imagine. Many electronic devices are included in this category. The products can be created by designers' creativity or defined by benchmarking. Benchmarking is a method for measuring a company's operations against the best companies both inside and outside of the industry (Dieter 2000). The current product is compared with the best known product and improvement is found from the comparison. CNs can be also defined by reverse engineering. A well known product is torn down and the design process of the product is inferred.

In Figure 1.4, CNs are defined at the beginning and marketing resides after all the engineering processes end. Thus, it may seem that the two items do not have common aspects. They actually have fairly close relationships. A product is sellable in the market when it satisfies CNs. Therefore, when CNs are defined, market analysis should follow.

2.2 Functional Requirements

Functional requirements (FRs) are defined in the functional domain in Figure 1.4. CNs are mapped into the functional domain and they are transformed to FRs which are defined by engineers. Definition of FRs is the first step of the designer's activity. It is generally made of engineering terminologies. However, they should cover the meanings of CNs. FRs express the functional conditions that a product has to have. When FRs are defined in an incorrect manner, one cannot design a good product. Therefore, definition of FRs is the most important process and this process has the most influence on the total design process.

It is not easy to define good FRs. A designer must be very careful on the decision of FRs. Engineers tend to spend little time on this process; however, they should spend the most time here. Brainstorming is an efficient way of thinking. In brainstorming, a group of people get together and communicate with ideas. A group of people can come up with a better idea than a single person.

It is extremely important to think functionally. In other words, definition of FRs should be emphasized. Many designers tend to think in the physical domain in Figure 1.4. They try to transform CNs to physical entities directly. In this case, they tend to use the physical parts of an existing product and a new idea is difficult to come up with. This habit makes design difficult. Again, **functional thinking** for definition of FRs should be kept in mind. When multiple FRs are defined, some aspects of one FR should not be included in others. They should be mutually exclusive.

Example 2.1 [FRs for a toaster]
We have the following CN:

CN : I want to bake a piece of bread conveniently.

Define FRs to satisfy the above CN.

Solution

The following FRs are defined:

FR_1 : Bake the bread by heat.

FR_2 : Hold the bread while heating.

FR_3 : Hold/Grab the bread by hand without touching the heat source.

Example 2.2 [FRs for a refrigerator] (Lee et al. 1994, Suh 2001)
Suppose we have the following CNs:

CN_1 : I need a device that can store food in a fresh manner.

CN_2 : I want to have ice anytime.

CN_3 : I want to store some food for a long time.

Define FRs to satisfy the above CNs

Solution

The top level FRs can be defined as follows:

FR_1 : Freeze food or water for long-term preservation.

FR_2 : Maintain food at a cold temperature for short-term preservation.

It is noted that an FR is defined by an imperative sentence. Different designers can make different FRs. It is crucial to define excellent FRs.

2.3 Design Parameters

FRs are materialized by definition of design parameters (DPs). DPs are physical objects that are defined in the physical domain in Figure 1.4. DPs are defined to satisfy FRs through a mapping process. As illustrated in Figure 1.4, FRs and DPs are defined in hierarchical structures. The hierarchy of the physical domain corresponds to that of the functional domain. DPs in a certain level are defined according to FRs in the same level and FRs in the next level are defined according to DPs in the previous level. Therefore, the mapping process is a zigzagging process. FRs and DPs of the top level are important in the overall design process because they have a lot of influence on the next levels. Generally, this FR-DP mapping process is called the design process.

Example 2.3 [DPs for the toaster of Example 2.1]

Define DPs that satisfy the FRs of Example 2.1.

Solution

The following DPs can be defined:

DP_1 : Electrical heating system

DP_2 : A mechanism to hold the bread while heating

DP_3 : A mechanism to draw out the bread

The product for the above DPs can be the one in Figure 2.1(a) or Figure 2.1(b). It is noted that a different product can be designed with the same DPs. That is, the final design depends on the designer's choice.

Example 2.4 [DPs for the refrigerator of Example 2.2]
Define appropriate DPs for the FRs of Example 2.2

Solution

DPs corresponding to the FRs are defined as follows:

DP_1 : The freezer section

DP_2 : The chiller section

2.4 Process Variables

DPs should be manufactured to produce a product. Process variables (PVs) are the variables that transform DPs into a real product. PVs reside in the process domain in Figure 1.4. The DP-PV mapping process is similar to that of the FR-DP mapping process. Although manufacturing is a fairly important factor in engineering, it is beyond the scope of this volume.

2.5 Constraints

When we perform some engineering activities, we usually have restrictions. Engineering activities should be carried out while these restrictions are satisfied. These restrictions are called constraints. In the design process, the constraints must be satisfied first. The constraints are defined by design specifications, current environments, customers, and so on. As illustrated in Figure 1.4, constraints reside in the physical and process domains. In Example 2.3, different designs are defined with the same DPs. One possible reason is existence of constraints. Based on the constraints, different designs can be generated.

(a) Toaster 1 (b) Toaster 2

Figure 2.1. Two toasters with the same DPs

2.6 Exercises

2.1 Consider a thermoplastic cup that can be used to dispense hot coffee from a vending machine. The coffee cup must be stiff enough so that it can be held by hand and also must have enough thermal insulation since it must be held by hand without burning the hand of the holder. The cost must also be low to be competitive with other products. What are the functional requirements (FRs)? Propose a design. How would you satisfy the FRs with your proposed design?

2.2 The government policy is to provide education to everyone who wants to receive a college education in Korea. At the same time, Korean people are eager to have one of the Korean universities to be in the top ten among all the universities in the world. State functional requirements (FRs). Design a system that can satisfy the FRs.

2.3 Suppose we want to design a missile that can seek in-coming enemy missiles and shoot them down. Our missile is of a cylindrical shape with a circular cross-section. The diameter of the missile should be as small as possible to minimize the drag on the missile. At the front end of the missile we want to put a radar system, which may be thought of as a 3 cm thick circular disk. The efficiency of the radar is proportional to the size of the radar and therefore, the diameter of radar should be made as large as possible. The radar must be oriented toward the enemy missile by turning it and controlling its orientation precisely.

a. What are the customer needs?
b. What are the functional requirements (FRs) for the mechanism that controls the orientation of the radar?
c. How would you satisfy the FRs?

2.4 We are thinking of creating a hospital at Ansan for our students, faculty, staff, and their families. What are the things we should consider in planning to develop an ideal Ansan hospital? What functional requirements (FRs) should the Ansan hospital satisfy to meet the majority of its customers? Please propose a design for the Ansan hospital.

2.5 We want to develop a desalination plant that can remove NaCl from seawater using solar energy. How should we do it? One idea is to buy an old ship that can no longer be used as a ship and take it to the tropic region off the coast of the central Africa and make it into a solar-powered desalination plant to supply water for agricultural purpose. What are the FRs for such a desalination plant? How can you satisfy the FRs?

2.6 In developing a new drug, it is necessary to test how the drug reacts with certain proteins in a cell. Cells are about 100 microns in diameter and the drug molecules are about 10 nanometers. Right now most pharmaceutical companies test the drug by sending it to outside of the cell and by watching how the biological system reacts to the new drug. It will be highly desirable to put the drug inside the cell and watch the reaction going on inside the cell. How would you do it? What are the FRs?

2.7 We want to create a computer game that a person can play to test his/her creativity. What should we do? What are the FRs? What are the design parameters? How would you measure the effectiveness of your game?

2.8 We want to use the tide of the Yellow Sea to generate electric power. Can you design a system that can generate electrical power from the ocean tide? What are the functional requirements (FRs) that we must satisfy? What are your design parameters?

2.9 The globe is warming up (*i.e.*, the global warming) due to the increase in the level of CO_2 in the atmosphere, which traps thermal energy by reducing the radiation of thermal energy from Earth to outer space. There have been a number of suggestions made on how to achieve the reduction of CO_2. How would you do it? What are functional requirements (FRs) that must be satisfied to control and reduce the amount of CO_2? How would you satisfy the FRs?

2.10 The transportation time by car across a large metropolitan city (a la Seoul) is a function of the size of the city. We may assume that the time it takes to go across the city is proportional to the square of the diameter of the city, which is equivalent to assuming that the traffic flow in a large city can be described by a diffusion model (a la thermal and mass diffusion). How would you design a new mega city to minimize the transportation time, which should also reduce the pollution level of the city? What are the FRs we need to satisfy? What are your DPs?

2.11 Most leading universities have a "need-blind" admissions policy, which states that all admitted students can study at the university regardless of whether or not they can pay the tuition. At these universities, the student is admitted based only on their academic achievements. After the student is accepted, then the school examines the ability of the parents to pay the student's tuition. Based on the financial assessment of the ability to pay, the school gives out scholarships to the student and the rest is paid by parents or the student. Your job is to devise a system for deciding on the amount of scholarship. What are the FRs and what are your DPs?

2.13 At University, we must promote friendship among all our students, including students from other countries. What are the issues? What FRs should we satisfy to achieve our goal? How would you organize the groups to achieve this goal?

2.14 Many people have tried to build a "human-powered flying machine." We want to design a flying machine. What are the FRs that we must satisfy? Propose your design that can satisfy the FRs (Please check your design against those reported in the literature and make sure that your design is unique).

2.15 When we transmit a message over a communications channel, the message can be corrupted because of the noise in the transmission line. Therefore, a continuing effort has been made to improve the quality of transmission. Define the FRs and devise a means of minimizing the error and maximizing the transmission of the information. (Now some people use the so-called the Turbo code to maximize the channel capacity, but you should think about your own solution.)

2.16 We transport liquefied natural gas using specially designed LNG ships that can contain such cold liquids. What are the FRs of such a ship? How would you design the thermal insulation system for such a ship?

2.17 One of the requirements of cell phones (hand phones) is the minimum use of battery power. Many different techniques have been devised to minimize the consumption of electrical energy. What are the FRs that must be satisfied to achieve this goal? Conceptually how would you design a system that will minimize the use of battery power?

2.18 A university student was flying on a Boeing 777 across the Pacific Ocean from Inchon, Korea, to San Francisco, U.S.A. It took roughly 11 hours. He noticed the following:

a. First hour of flying: Altitude = 33,000 ft, Ground speed 610 mph
b. Third hour of flying: Altitude = 34,000 ft, Ground speed 620 mph
c. Sixth hour of flying: Altitude = 35,000 ft. Ground speed 630 mph
d. Tenth hour of flying: Altitude = 39,000 ft, Ground speed 610 mph

On his return trip to Korea from the United States, which took 13 hours, th e student noticed the following from the flight monitor:

e. First hour of flying: Altitude = 33,000 ft, Ground speed 550 mph
f. Third hour of flying: Altitude = 34,000 ft, Ground speed 560 mph
g. Sixth hour of flying: Altitude = 35,000 ft. Ground speed 540 mph
h. Tenth hour of flying: Altitude = 39,000 ft, Ground speed 540 mph

From the data given above, can you explain, with a first order approximation, the design of the aircraft? How do you determine the engine size? How do you determine the wing size? How would you improve the fuel efficiency of the airplane?

2.19 There are the Internet-based search engines such as Google and Naver, which extract relevant reference materials when keywords are provided by the user. The next generation of the advanced search engines should extract relevant information from selected references and synthesize the answer for the user based on the background-information on the user of the software, *e.g.*, educational level, field of specialization, experience, *etc.* What are the FRs of such a software system? Design the software system. What are the DPs of your software system? You do not have to code the software.

2.20 Korea needs to develop knowledge-based industries that can be the engine for its economic growth in 20 years. We can learn a lot about the prerequisites that must be satisfied to develop high-tech industries by examining the history of high-tech centers in the United States. State FRs for developing knowledge-based industries and design a national policy for satisfying the FRs. What are your DPs?

III. Axiomatic Design

3.1 Introduction

Axiomatic design is a design methodology that was created and popularized by Professor Suh of the Massachusetts Institute of Technology (Suh 1990, 2001). It is a design framework that works on all the design disciplines. Axiomatic design consists of two axioms: the Independence Axiom and the Information Axiom. A good design should satisfy the two axioms.

An axiom is a statement accepted without proof as an underlying assumption of a formal mathematical theory. It is also called a postulate. It cannot be mathematically proved. If a counter example is found, the axiom becomes obsolete. Axioms are mostly used in (Euclidean) geometry. Geometry starts with axioms, and theorems and corollaries are derived from the axioms.

Points and lines are names for the elements of two (distinct) sets. Incidence is a relationship that may (or may not) hold between a particular point and a particular line. The followings are examples of axioms:

(1) For every two points, there exists a line incident with both points.
(2) For every two points, there is no more than one line incident with both points.
(3) There exist at least two points incident with each line.
(4) There exist at least three points. Not all points are incident with the same line.

Also, many laws in physics are axioms as well. For example, Newton's laws are axioms. The following Newton's equation is an axiom:

$$F = ma \tag{3.1}$$

where F is the external force, m is the mass and a is the acceleration. Many other physics laws (theorems) are made from Newton's laws. Thermodynamic principles are also axioms. We cannot prove them; however, we cannot find a machine that violates the thermodynamic principles.

3.2 Design Axioms

As mentioned earlier, there are two design axioms and they are as follows:

Axiom 1: The Independence Axiom

Maintain the independence of FRs
Alternate Statement 1: An optimal design always maintains the independence of FRs.
Alternate Statement 2: In an acceptable design, DPs and FRs are related in such a way that a specific DP can be adjusted to satisfy its corresponding FR without affecting other functional requirements.

Axiom 2: The Information Axiom

Minimize the information content of the design
Alternate Statement: The best design is a functionally uncoupled design that has minimum information content.

The axioms are utilized sequentially. First, a design is found using the Independence Axiom. If multiple designs that satisfy the Independence Axiom are found, the Information Axiom is used to find the best one. A flow chart on how to use the axioms is illustrated in Figure 3.1. This will be explained later.

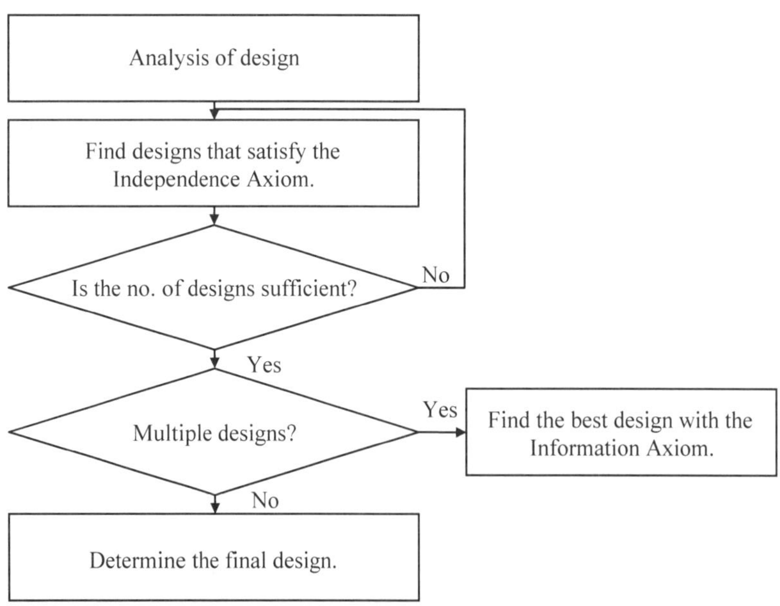

Figure 3.1. Flow chart of the application of axiomatic design

3.3 The Independence Axiom

The Independence Axiom indicates that the relationship between FRs and DPs should be independent. That is, a DP should satisfy a corresponding FR. Suppose we have three FRs and three DPs. If we use the vector notation

$$\mathbf{FR} = \begin{bmatrix} FR_1 \\ FR_2 \\ FR_3 \end{bmatrix}, \qquad \mathbf{DP} = \begin{bmatrix} DP_1 \\ DP_2 \\ DP_3 \end{bmatrix} \tag{3.2}$$

Using design matrix **A**, the relationship (design equation) of **FR** and **DP** is as follows:

$$\mathbf{FR} = \mathbf{A}\mathbf{DP} \tag{3.3a}$$

$$\begin{bmatrix} FR_1 \\ FR_2 \\ FR_3 \end{bmatrix} = \begin{bmatrix} A_{11} & A_{12} & A_{13} \\ A_{21} & A_{22} & A_{23} \\ A_{31} & A_{32} & A_{33} \end{bmatrix} \begin{bmatrix} DP_1 \\ DP_2 \\ DP_3 \end{bmatrix} \tag{3.3b}$$

FR-DP relationships according to matrix **A** are shown in Table 3.1. If the design matrix is a diagonal matrix as shown in the first case of Table 3.1, it is an uncoupled design. Because each DP can satisfy a corresponding FR, the uncoupled design perfectly satisfies the Independence Axiom. When the design matrix is triangular as shown in the second case of Table 3.1, the design is a decoupled design. A decoupled design satisfies the Independence Axiom if the design sequence is correct. In the second case of Table 3.1, DP_1 is first determined for FR_1 and fixed. FR_2 is satisfied by the choice of DP_2 and the fixed DP_1. DP_3 is determined in the same manner with the fixed DP_1 and DP_2.

When a design matrix is neither diagonal nor triangular, the design becomes a coupled design. In a coupled design, no sequences of DPs can satisfy the FRs independently. Therefore, an uncoupled or a decoupled design satisfies the Independence Axiom and a coupled design does not. If a design is coupled, an uncoupled or decoupled design must be found through a new choice of DPs. It is noted that when the numbers of FRs and DPs are different the design is coupled. For the *i*th FR or DP, the subscript notation is used in this book. FR_i is frequently expressed by *FRi*. With design matrices, multiplication and addition are permitted; however, other manipulations such as coordinate transformation are not permitted.

Table 3.1. FR-DP relationship according to the design matrix

	Design equation	Design process
Uncoupled design	$\begin{bmatrix} FR_1 \\ FR_2 \\ FR_3 \end{bmatrix} = \begin{bmatrix} A_{11} & 0 & 0 \\ 0 & A_{22} & 0 \\ 0 & 0 & A_{33} \end{bmatrix} \begin{bmatrix} DP_1 \\ DP_2 \\ DP_3 \end{bmatrix}$	$FR_1 = A_{11} \times DP_1$ $FR_2 = A_{22} \times DP_2$ $FR_3 = A_{33} \times DP_3$
Decoupled design	$\begin{bmatrix} FR_1 \\ FR_2 \\ FR_3 \end{bmatrix} = \begin{bmatrix} A_{11} & 0 & 0 \\ A_{21} & A_{22} & 0 \\ A_{31} & A_{32} & A_{33} \end{bmatrix} \begin{bmatrix} DP_1 \\ DP_2 \\ DP_3 \end{bmatrix}$	$FR_1 = A_{11} \times DP_1$ $FR_2 = A_{21} \times DP_1 + A_{22} \times DP_2$ $FR_3 = A_{31} \times DP_1 + A_{32} \times DP_2$ $+A_{33} \times DP_3$
Coupled design	$\begin{bmatrix} FR_1 \\ FR_2 \\ FR_3 \end{bmatrix} = \begin{bmatrix} A_{11} & A_{12} & A_{13} \\ A_{21} & A_{22} & A_{23} \\ A_{31} & A_{32} & A_{33} \end{bmatrix} \begin{bmatrix} DP_1 \\ DP_2 \\ DP_3 \end{bmatrix}$	$FR_1 = A_{11} \times DP_1 + A_{12} \times DP_2$ $+A_{13} \times DP_3$ $FR_2 = A_{21} \times DP_1 + A_{22} \times DP_2$ $+A_{23} \times DP_3$ $FR_3 = A_{31} \times DP_1 + A_{32} \times DP_2$ $+A_{33} \times DP_3$

It is noted that constraints (Cs) exist in the design. Constraints are generally defined from design specifications and they must be satisfied. Constraints can be defined without regard to the independence of FRs and coupled by DPs. As illustrated in Figure 1.4, the constraints can be defined in the DP or PV domains.

3.4 Application of the Independence Axiom

The following examples show applications of the Independence Axiom. As mentioned earlier, an imperative sentence is used for the expression of an FR and a noun is used for a DP.

Example 3.1 [Design of a Refrigerator Door] (NSF 1998, Suh 2001)
Figure 3.2 shows two refrigerator doors that we most frequently encounter. Which one has the better design? To answer the question, the doors are analyzed based on the axiomatic design viewpoint. Functional requirements are defined as follows:

FR_1 : Provide access to the items stored in the refrigerator.

FR_2 : Minimize energy loss.

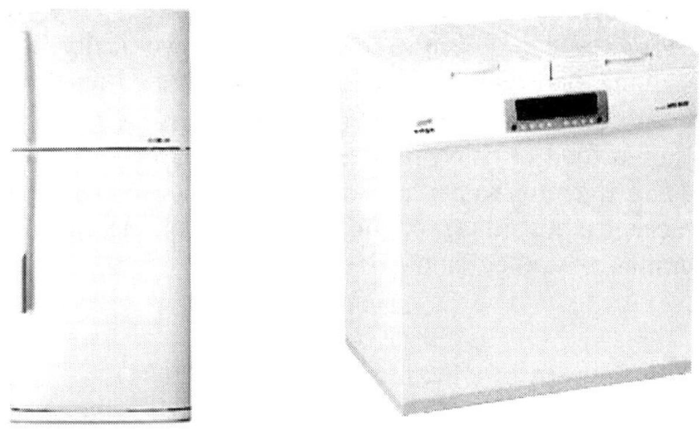

(a) Vertically hung door (b) Horizontally hung door

Figure 3.2. Refrigerator

Solution

Design parameters for the vertically hung door in Figure 3.2(a) are as follows:

DP_1 : Vertically hung door

DP_2 : Thermal insulation material in the door

The design equation may be stated as

$$\begin{bmatrix} FR_1 \\ FR_2 \end{bmatrix} = \begin{bmatrix} X & 0 \\ X & X \end{bmatrix} \begin{bmatrix} DP_1 \\ DP_2 \end{bmatrix} \tag{3.4}$$

where X indicates a nonzero value, and hence a dependence between an FR and a DP.

The design in Equation 3.4 is a decoupled one and satisfies the Independence Axiom. However, when we open the door, energy loss occurs due to the X in the off-diagonal term. Now, the horizontally hung door in Figure 3.2(b) is analyzed.

DP_1 : Horizontally hung door

DP_2 : Thermal insulation material in the door

The design equation is made as follows:

$$\begin{bmatrix} FR_1 \\ FR_2 \end{bmatrix} = \begin{bmatrix} X & 0 \\ 0 & X \end{bmatrix} \begin{bmatrix} DP_1 \\ DP_2 \end{bmatrix} \tag{3.5}$$

When we open the horizontally hung door, cold air remains in the refrigerator and energy loss can be minimized. Therefore, the horizontally hung door has an uncoupled design and is a better design than the vertically hung door. Is the horizontally hung door always better? As far as the functional requirements defined here are kept, it is correct. Suppose that constraints are proposed for the amount of stored food or convenience to access items. Then the problem will be different. If a refrigerator with a horizontally hung door violates the constraints, it cannot be accepted regardless of the satisfaction of the Independence Axiom. When constraints exist, they should be checked first.

Example 3.2 [Design of a Water Faucet] (Suh 2001)

A faucet is designed. The user should be able to control the temperature and the running rate of water. Since there are many commercialized faucets, they are evaluated. The functional requirements of a faucet are defined as follows:

FR_1 : Control the flow of water (Q).

FR_2 : Control the temperature of water (T).

Solution

Analyzing the product in Figure 3.3(a), DPs and the design equation are defined as follows:

DP_1 : Angle ϕ_1

DP_2 : Angle ϕ_2

$$\begin{bmatrix} FR_1(Q) \\ FR_2(T) \end{bmatrix} = \begin{bmatrix} X & X \\ X & X \end{bmatrix} \begin{bmatrix} DP_1(\phi_1) \\ DP_2(\phi_2) \end{bmatrix} \tag{3.6}$$

As shown in Equation 3.6, the design is coupled. Thus, the design is not acceptable.

Another example is presented in Figure 3.3(b). The design is analyzed as follows:

DP_1 : Angle ϕ_1

DP_2 : Angle ϕ_2

$$\begin{bmatrix} FR_1(Q) \\ FR_2(T) \end{bmatrix} = \begin{bmatrix} X & 0 \\ 0 & X \end{bmatrix} \begin{bmatrix} DP_1(\phi_1) \\ DP_2(\phi_2) \end{bmatrix} \tag{3.7}$$

Because the design matrix is diagonal, the design is uncoupled. Therefore, it satisfies the Independence Axiom and is acceptable.

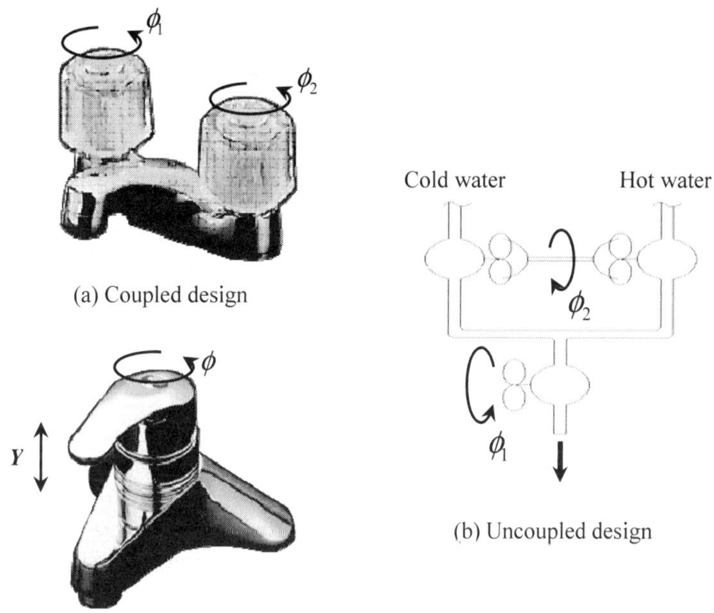

(a) Coupled design

(b) Uncoupled design

(c) Uncoupled design

Figure 3.3. Example of a water faucet

One more design is illustrated in Figure 3.3(c).

DP_1 : Displacement Y
DP_2 : Angle ϕ

$$\begin{bmatrix} FR_1(Q) \\ FR_2(T) \end{bmatrix} = \begin{bmatrix} X & 0 \\ 0 & X \end{bmatrix} \begin{bmatrix} DP_1(Y) \\ DP_2(\phi) \end{bmatrix} \tag{3.8}$$

The design matrix is diagonal; therefore, the design is uncoupled. We have two uncoupled designs. Which one is better? It is easy to manipulate the one in Figure 3.3(c). This can be explained by the Information Axiom which will be introduced later. The design in Figure 3.3(c) is the best from the viewpoint of the Information Axiom. Actually, the one in Figure 3.3(c) is becoming popular. This conclusion is made based on engineering functional requirements. If aesthetic aspects are important, different decisions can be made.

Example 3.3 [Axiomatic Design of the Toaster in Example 2.1]

Using the FRs of Example 2.1 and DPs of Example 2.3, make the design matrices for the products in Figure 2.1(a) and Figure 2.1(b).

Solution

The design matrix for each design is defined as follows:

$$\begin{bmatrix} FR_1 \\ FR_2 \\ FR_3 \end{bmatrix} = \begin{bmatrix} X & 0 & 0 \\ 0 & X & x \\ 0 & X & X \end{bmatrix} = \begin{bmatrix} DP_1 \\ DP_2 \\ DP_3 \end{bmatrix} \quad \text{(for Figure 2.1(a))} \tag{3.9a}$$

$$\begin{bmatrix} FR_1 \\ FR_2 \\ FR_3 \end{bmatrix} = \begin{bmatrix} X & 0 & 0 \\ 0 & X & x \\ 0 & X & X \end{bmatrix} = \begin{bmatrix} DP_1 \\ DP_2 \\ DP_3 \end{bmatrix} \quad \text{(for Figure 2.1(b))} \tag{3.9b}$$

When we design a complicated system, a definition of a simple FR-DP relationship may not be sufficient. Then we can decompose the relationship. As illustrated in Figure 3.4, a new relationship is defined by the zigzagging process between the functional and physical domains. The zigzagging process is presented by the numbers in Figure 3.4. It is noted that DPs are defined according to FRs in the same level and FRs of the lower level are defined based on the characteristics of DPs in the upper level. This decomposition process continues until the leaf (bottom) level is reached.

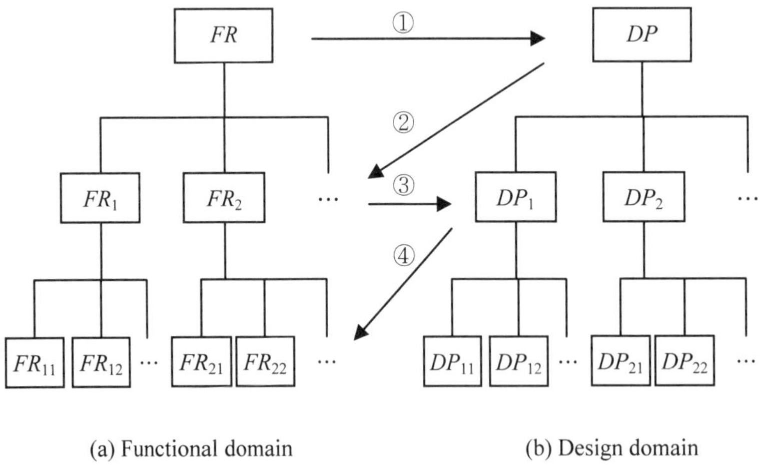

Figure 3.4. Zigzagging process between domains

Example 3.4 [Decomposition of Example 2.2]

Decompose the FRs of Example 2.2 and design the refrigerator by using the axiomatic approach.

The design matrix for Example 2.2 (first level) is diagonal; therefore, it is an uncoupled design. FR_1 can be decomposed by the selection of DP_1.

FR_{11} : Maintain the temperature of the freezer section in the range of $-18°C \pm 2°C$.

FR_{12} : Maintain a uniform temperature in the freezer section.

FR_{13} : Control the relative humidity to 50% in the freezer section.

In the same manner, FR_2 can be decomposed with respect to DP_2.

FR_{21} : Maintain the temperature of the chiller section in the range of $2°C - 3°C$.

FR_{22} : Maintain a uniform temperature in the chiller section within $\pm 0.5°C$ of the preset temperature.

The design parameters for the second level are to be determined. The DPs must be determined to satisfy the independence of the FRs. It is noted that DPs in the lower level should be determined so as not to violate the independence of the upper level.

The FRs of the freezer section can be satisfied by (1) a device pumping chilled air into the freezer section, (2) a device for circulation of air for a uniform temperature, (3) a monitoring device to independently control the temperature and humidity. Therefore, the DPs in the second level are defined as follows:

DP_{11} : Sensor/compressor system that activates the compressor when the temperature of the freezer section is different from the preset one

DP_{12} : Air circulation system that blows the air into the freezer and circulates it uniformly

DP_{13} : Condenser that condenses the moisture in the returned air when the dew point is exceeded

The design is a decoupled one as follows:

$$\begin{bmatrix} FR_{12} \\ FR_{11} \\ FR_{13} \end{bmatrix} = \begin{bmatrix} X & 0 & 0 \\ X & X & 0 \\ X & 0 & X \end{bmatrix} \begin{bmatrix} DP_{12} \\ DP_{11} \\ DP_{13} \end{bmatrix} \tag{3.10}$$

For food storage in the chiller section, the temperature should be maintained in the range of $2°C - 3°C$. The chiller section also activates the compressor and circulates the air. Design parameters for the chiller section are

DP_{21}: Sensor/compressor system that activates the compressor when the temperature of the chiller section is different from the preset one

DP_{22}: Air circulation system that blows the air into the chiller section and circulates it uniformly

The design equation is a decoupled one as follows:

$$\begin{bmatrix} FR_{22} \\ FR_{21} \end{bmatrix} = \begin{bmatrix} X & 0 \\ X & X \end{bmatrix} \begin{bmatrix} DP_{22} \\ DP_{21} \end{bmatrix} \tag{3.11}$$

The entire design equation decomposed up to the second level is a decoupled one as follows:

$$\begin{bmatrix} FR_{12} \\ FR_{11} \\ FR_{13} \\ FR_{22} \\ FR_{21} \end{bmatrix} = \begin{bmatrix} X & 0 & 0 & 0 & 0 \\ X & X & 0 & 0 & 0 \\ X & 0 & X & 0 & 0 \\ 0 & 0 & 0 & X & 0 \\ 0 & 0 & 0 & X & X \end{bmatrix} \begin{bmatrix} DP_{12} \\ DP_{11} \\ DP_{13} \\ DP_{22} \\ DP_{21} \end{bmatrix} \tag{3.12}$$

It is noted that the FRs of the lower level still keep the independence of the upper level in Equation 3.12.

From the design equation in Equation 3.12, one compressor and two fans can satisfy the FRs. DP_{11} and DP_{21} are sensor/compressor systems so that the compressor is activated by the sensors. However, the fans of DP_{12} and DP_{22} will not be activated unless the temperature is out of the range of the preset one. Therefore, the design with one compressor and two fans satisfies the Independence Axiom. An example is illustrated in Figure 3.5. Other designs can be proposed. If multiple designs are proposed, we can select one that satisfies the Independence Axiom and controls the temperature and humidity in a wide range.

The new design and the conventional refrigerator are compared. The conventional refrigerator consists of one compressor and one fan. As illustrated in Figure 3.6, a damper is utilized to cool the refrigerating room. Therefore, the temperature of the refrigerator is not independently controlled. When the temperature exceeds $3°C$, the damper is opened. However FR_{21} is not satisfied unless the compressor and the fan of the freezer section are activated.

There is a saying that a simple design is a good one. From this statement, we may guess that a good design makes one DP satisfy multiple FRs. In other words, a coupled design is better. This aspect is very confusing in axiomatic design. However, from an axiomatic design viewpoint, this is the case where multiple DPs make a physical entity. That is, a physical entity consists of multiple DPs and

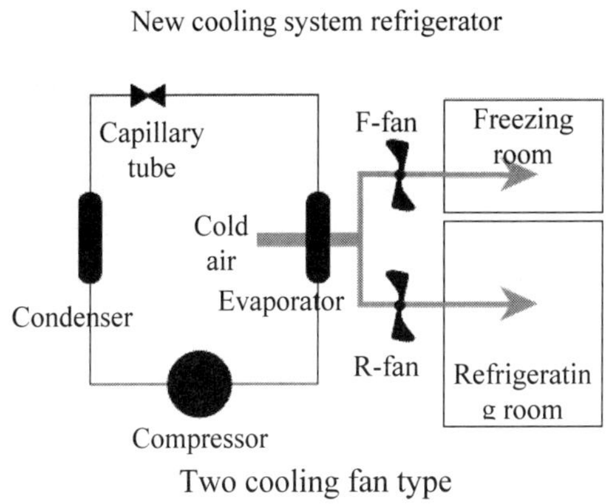

Figure 3.5. A new design of a refrigerator that satisfies the Independence Axiom

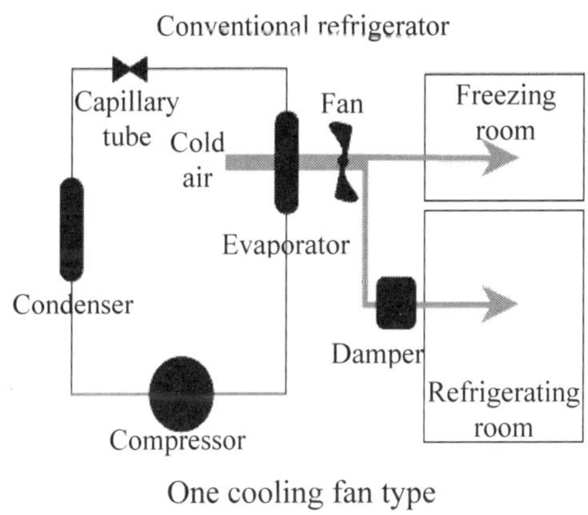

Figure 3.6. Conventional refrigerator

multiple DPs satisfy FRs of the same number. This is called "physical integration." Physical integration is desirable because the information quantity can be reduced. The following example is a typical example of physical integration.

Example 3.5 [Bottle-can Opener] (NSF 1998, Suh 2001)

Suppose we need a device that can open bottles and cans. Functional requirements are defined as follows:

FR_1 : Design a device that can open bottles.

FR_2 : Design a device that can open cans.

Solution

Figure 3.7. Bottle-can opener

Figure 3.8. Beverage can

The device in Figure 3.7 has one physical entity for the bottle opener and can opener. However, two DPs at both ends independently satisfy the two functional requirements. Therefore, the design in Figure 3.7 satisfies the Independence Axiom. If the constraint set includes "both functions should be simultaneously used," then a different design should be investigated.

Example 3.6 [Beverage Can Design] (NSF 1998, Suh 2001)
Consider an aluminum beverage can that contains liquid as illustrated in Figure 3.8. According to an expert working at one aluminum can manufacturer, there are 12 FRs for the can. Plausible FRs: contain axial and radial pressure, withstand moderate impact when the can is dropped from a certain height, allow stacking on top of each other, provide easy access to the liquid in the can, minimize the use of aluminum, be printable on the surface, and more. However, these 12 FRs are not satisfied by 12 physical pieces. The can consists of three pieces: the body, the lid and the tab opener. There must be at least 12 DPs and they are distributed to these three pieces. Most of the DPs are associated with the geometry of the can: the thickness of the body, the curvatures at the bottom, the reduced diameter at the top to reduce the material used to make the top lid, the corrugated geometry of the tab opener to increase the stiffness, the small extrusion on the lid to attach the tab, *etc.*

The complexity is reduced when physical integration is utilized while the independence is maintained. That is, related information quantity is reduced. Therefore, physical integration does not violate the Independence Axiom. Instead, it is recommended.

3.5 The Information Axiom

Axiomatic design requires satisfaction of the Independence Axiom. Multiple designs that satisfy the Independence Axiom can be derived. In this case, the best design should be selected. The best design is the one with minimum information. How can we quantitatively define the information content? The definition varies according to the situation. Generally, the information is related to complexity. Then how can we measure complexity? We need a rigorous definition for the information content. The information content can be differently defined according to the characteristics of the design. The probability of success has been utilized as an index of the information content.

Suppose p is the probability of satisfying FR with DP . Then the information content is defined as

$$I = \log_2 1/p \tag{3.13}$$

In Equation 3.13, the reciprocal of p is used to make the larger probability have less information. Also, the logarithm function is utilized to enhance additivity. The base of the logarithm is 2 to express the information content with the bit unit.

Suppose we have the following uncoupled design:

$$\begin{bmatrix} FR_1 \\ FR_2 \\ FR_3 \end{bmatrix} = \begin{bmatrix} A_{11} & 0 & 0 \\ 0 & A_{22} & 0 \\ 0 & 0 & A_{33} \end{bmatrix} \begin{bmatrix} DP_1 \\ DP_2 \\ DP_3 \end{bmatrix} \tag{3.14}$$

Suppose p_1, p_2 and p_3 are the probabilities of satisfying FR_1 , FR_2 and FR_3 with DP_1, DP_2 and DP_3, respectively. The total information I_{total} is

$$I_{\text{total}} = \sum_{i=1}^{3} I_i = \sum_{i=1}^{3} \log_2\left(\frac{1}{p_i}\right) \tag{3.15}$$

It is noted that the information content should only be defined based on the corresponding functional requirement.

Calculation of the information content for a decoupled design is somewhat different. Since independence is satisfied by the sequence of the process, the probability of success of the later process depends on that of the previous one. Therefore, it is a conditional probability. Suppose we have the following decoupled design:

$$\begin{bmatrix} FR_1 \\ FR_2 \end{bmatrix} = \begin{bmatrix} X & 0 \\ X & X \end{bmatrix} \begin{bmatrix} DP_1 \\ DP_2 \end{bmatrix} \tag{3.16}$$

If p_1 is the probability that DP_1 satisfies FR_1, then the probability that DP_2 satisfies FR_2 under the satisfaction of FR_1 by DP_1 is a conditional probability. Suppose it is p_{21}. Then the probability of success p that both FR_1 and FR_2 are satisfied is

$$p = p_1 p_{21} \tag{3.17}$$

The total information content for p is

$$I = -\log_2 p = -\log_2(p_1 p_{21}) = -\log_2 p_1 - \log_2 p_{21} = I_1 + I_2 \tag{3.18}$$

The conditional probability is useful for investigating the characteristics of the Information Axiom. However, it is rarely applied to real problems because p_{21} is not easy to evaluate. Instead, the probability density function is more practical for application.

Information content can be calculated by using the probability density function. Figure 3.9 presents a schematic view of this. The terminologies are as follows: the design range is the range for the design target, the system range is the operating range of the designed product and the common range is the common area between the design range and the system range. The design range is defined by lower and upper bounds and the system range is defined by a distribution function of the system performance. A uniform distribution of a system range is illustrated in

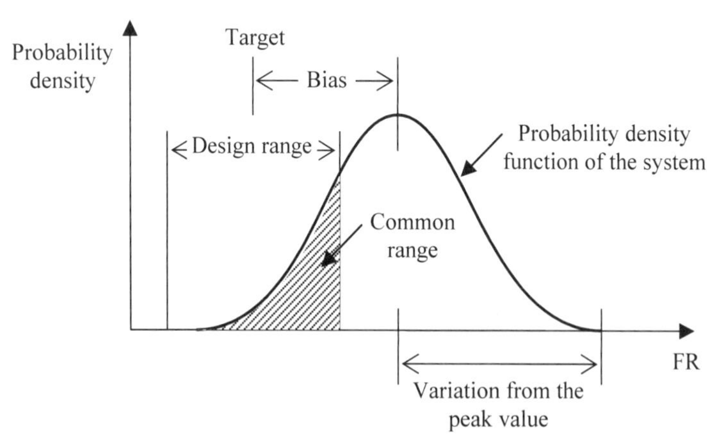

Figure 3.9. Calculation of the information content using the probability density function

Figure 3.10. The design should be directed to increase the common range. The information content is defined as follows:

$$p_s = A_{cr} / A_{sr} \tag{3.19a}$$

$$I = -\log_2(A_{cr} / A_{sr}) \tag{3.19b}$$

where A_{sr} is the system range and A_{cr} is the common range.

3.6 Application of the Information Axiom

Example 3.7 [An Example of Calculating Information Content]
Information content is calculated for the design problem in Figure 3.7. It is assumed that the probability of satisfying FR_1 with DP_1 is 0.9 and the one for FR_2 with DP_2 is 0.85. The total information content is as follows:

$$I_{total} = I_1 + I_2 = \log_2\left(\frac{1}{0.9}\right) + \log_2\left(\frac{1}{0.85}\right) = 0.1520 + 0.2345 = 0.3865\,(\text{bits}) \tag{3.20}$$

Now the reduction of information due to physical integration is explained with Example 3.5. Without physical integration, two pieces of the two DPs should be made. If we keep the amount of material constant, the sizes of each piece should be smaller. Then the use of each piece is inconvenient and the probability of success is reduced. The result is that the information content is increased. Therefore, it is inferred that a tool with physical integration has less information content. However, not much research has been done on quantifying the reduction of information content from physical integration. We need more research on this topic.

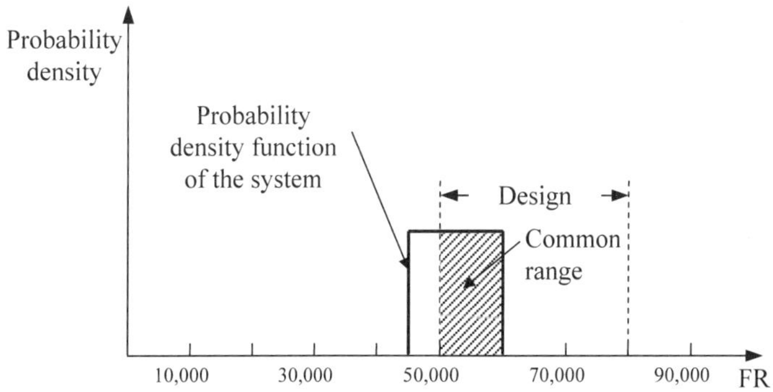

Figure 3.10. Probability density function of a uniform distribution

Example 3.8 [Manufacturing a Bar with a Specified Tolerance]

Another method to calculate the probability of success is introduced. A bar of $1\,\text{m}$ length is to be manufactured. The tolerances of two cases are $\pm 0.00001\,\text{m}$ and $\pm 0.1\,\text{m}$. Calculate the information content for both cases.

Solution

If we use the same machine for both cases, the probability of success is smaller when the tolerance is small. Also, if the given length (nominal length) is longer, the ratio of the tolerance to the total length is smaller. Thus, the probability of success is as follows:

$$p = f\left(\frac{\text{tolerance}}{\text{nominal length}}\right) \tag{3.21}$$

If we assume that Equation 3.21 is linear, then it becomes as follows:

$$p = c\frac{\text{tolerance}}{\text{nominal length}} \tag{3.22}$$

where c is a constant.

Example 3.9 [Calculation of the Information Content Using the Probability Density Function]

A problem is artificially made to demonstrate an example. A person defines two functional requirements to buy a house as follows:

FR_1 : Let the price range be from 50,000 dollars to 80,000 dollars.

FR_2 : Let the commuting time be within 40 minutes.

The person considers a house in city A or city B. Table 3.2 shows the conditions of both cities. Where should the person buy a house to minimize the information content?

Solution

The system range is defined from Table 3.2 and the design range is determined from the functional requirements. It is assumed that all the probability densities are uniform. Figure 3.10 presents the probability density for the price of the house in city A. Other items can be illustrated in the same manner. The information content for city A is as follows:

Table 3.2. Conditions for each city

	city A	city B
Price	\$45,000-\$60,000	\$70,000-\$90,000
Commuting time	35-50 min	20-30 min

$$I_{A1} = \log_2\left(\frac{1.5}{1}\right) = 0.59, \quad I_{A2} = \log_2\left(\frac{15}{5}\right) = 1.59 \tag{3.23}$$

$$I_A = I_{A1} + I_{A2} = 2.18\ \text{(bits)} \tag{3.24}$$

In the same manner, the information content for city B is

$$I_{B1} = \log_2\left(\frac{2}{1}\right) = 1.0\,, \quad I_{B1} = \log_2\left(\frac{10}{10}\right) = 0.0 \tag{3.25}$$

$$I_B = I_{B1} + I_{B2} = 1.0\ \text{(bits)} \tag{3.26}$$

The information content I_A for city A is 2.18 and that for city B (I_B) is 1.0. Therefore, city B has the optimum house from an axiomatic design viewpoint.

The design should be directed to reduce the information content in Equation 3.13. From Figure 3.9, it is effective to reduce the bias that is the difference between the averages of the system range and design range. After that, the standard deviation of the system range should be decreased. Then the common range is increased and the information content is reduced. This aspect is related to robust design.

3.7 Discussion

As explained earlier, the two axioms are independent of each other. Thus, we have to apply them separately. Generally, the Independence Axiom should be satisfied first. In many cases, the design is terminated only with the application of the Independence Axiom. When both axioms are utilized, the flow in Figure 3.1 is recommended.

When we apply the Independence Axiom, the ideal design should be kept in mind. The numbers of FRs and DPs are the same in an ideal design. The design matrix should be a square diagonal or triangular one. If the numbers are different, the design is coupled. When the number of DPs is smaller, new DPs should be added. In a redundant design where the number of DPs is larger, the number of DPs should be reduced or some specific DPs should be fixed.

Suppose we have the following redundant design:

$$\begin{bmatrix} FR_1 \\ FR_2 \end{bmatrix} = \begin{bmatrix} X & X & 0 \\ 0 & X & X \end{bmatrix} \begin{bmatrix} DP_1 \\ DP_2 \\ DP_3 \end{bmatrix} \tag{3.27}$$

First, we can fix DP_2. Then the design becomes an uncoupled one. That is, redundant parameters are fixed to make the design uncoupled or decoupled with the rest of the parameters.

The Information Axiom is utilized to quantitatively evaluate a design that satisfies the Independence Axiom. It is especially useful when multiple designs are compared. When multiple designs, which satisfy the Independence Axiom, are found, the one with the minimum information content is selected as the final design.

Basically, axiomatic design can be exploited in creating a new design or in evaluating existing designs. It is quite useful in the conceptual design of new products. Although the history of the method is relatively short, its usefulness has been verified through many examples. There are some common responses from application designers. First, they tend to easily agree with the axioms and think that they can use them right away. However, they have difficulties in testing the axioms with their existing products. In most cases, they tend to look at the designs with previous concepts, not from an axiomatic viewpoint. Many designers stop applying axiomatic design at this stage. However, if the designers overcome this stage, they can realize the usefulness of axiomatic design. It is important not to consider existing products when the functional requirements are defined. Instead, designers should think about the functional requirements in a solution neutral environment. In recent research, axiomatic design is utilized in detailed designs. Later examples will demonstrate how axiomatic design is applied to the detailed design process of structures.

3.8 Exercises

3.1 Analyze the design of a mechanical pencil with the Independence Axiom.

3.2 Analyze the design of an electrical pencil sharpener with the Independence Axiom.

3.3 Perform the decomposition of Example 3.3 (the one in Figure 2.1(a)) with the axiomatic approach.

3.4 Find the design for Exercise 2.1 with the Independence Axiom.

3.5 Find the design for Exercise 2.4 with the Independence Axiom.

3.6 Find the design for Exercise 2.10 with the Independence Axiom.

3.7 Find the design for Exercise 2.17 with the Independence Axiom.

3.8 Find a product that uses the idea of physical integration and analyze it with the Independence Axiom.

3.9 Make a plan for your life with axiomatic design (use the zigzagging process).

IV. Design Project

4.1 About the Design Project

A project is a unique sequence of work tasks, undertaken once, to achieve a specific set of objectives (Eggert 2005). The word "project" is frequently used everywhere these days. In a design class, projects are usually given to students as the final large-scale homework and they are called design projects. A design project is a project where a design activity is performed. The design project is different from general homework or an exam. First, the scale of the design project is large and the problem usually has an open-ended solution. In other words, there can be many solutions or finding one solution can be quite difficult. The design project is carried out for a long period of time. Generally, half the semester can be spent to finish the design project. The flow of the design project is illustrated in Figure 4.1. It should be remembered that each step of Figure 4.1 should be explained reasonably.

In general, 3 to 5 students collaborate for a design project. They can design a product together. As mentioned earlier they have to be able to logically explain each step of their progress. The designed product can be manufactured. However, manufacturing the product may not be accomplished in a semester when the product is a large-scale one. In that case, only the design is sufficient.

4.2 Elements of a Design Project

In general, a project has three elements as illustrated in Figure 4.2. They are a client, a designer and a user (Dym and Little 2004). The client is the person who requests the project to be carried out. In Figure 1.4, the client defines CNs in the customer domain. The designer designs the product based on the CNs. A designer performs the other activities in Figure 1.4 except for the customer domain. The user is the final customer of the designed product. Actually, the client defines CNs to sell the product to the user. Therefore, the planning stage in Figure 1.1 has a close relationship with marketing.

We need various resources for a design project. They are design topics, students, supervisors (coaches), budget, and so on. First, topics are given. The topics can be defined by the supervisors of the class or students. In this case the client can be supervisors or students. The topics should be large enough for a team of students to work for weeks and small enough for the team to finish during the academic term. When the topics are defined by supervisors, they are presented to students. The students form teams and the team picks a topic. A team has the role of the designer in Figure 4.2. If the topic is defined by a group of students, they can form a team. In a special case, a single student can carry out a design project.

A group of students can form a team based on the selected topic. A single lecturer can manage all the teams or multiple supervisors may manage the teams as coaches. A supervisor can be a professor, a teaching assistant or both. The team follows the process as illustrated in Figure 4.1. The team should have meetings periodically. The supervisor checks the progress and advises on the direction of the work. It is recommended that a team has one supervisor. When multiple supervisors manage a team, the directions from the supervisors can be different

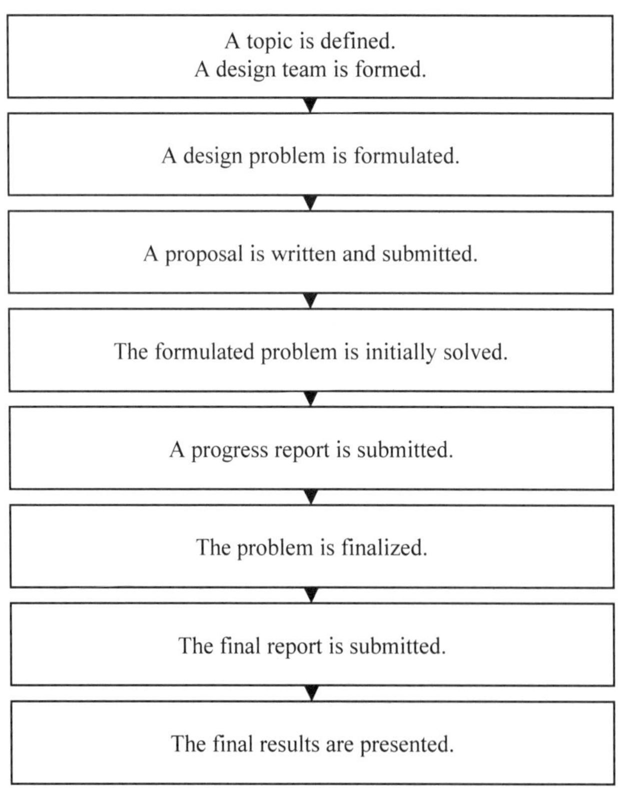

Figure 4.1. The flow of a design project

and the students may become confused. Sometimes, we need a budget for a design project. Especially, when manufacturing is required, the budget is important. Therefore, the course developer should secure an appropriate budget. As illustrated in Figure 4.1, the final report is submitted to the supervisor and the results are presented in the class. Thus, the user in Figure 4.2 is the supervisor and the audience of the presentation. It is most important to convey the results of the design process. The client is likely uninterested in the history of the project or in the design team's internal workings (Dym and Little 2004).

4.3 Team Project and Communication Skills

As the supervisor-to-students relationship is essential, the student-to-student relationship is also important. The entire job should be fairly divided among members. It is recommended that a leader of the team be selected. The leader manages the activities of the members. The leader can define the scope of responsibility for each member in the team to avoid duplication of effort. S/he can also establish the quality of workmanship and standard of presentation (Borowick 2000). The leader can call for meetings, check the overall progress and modulate the opinions of the team members.

Collaboration is crucial in the success of a design project. Members collaborate through communication. Therefore, verbal communication is central in the design process. Communication skills can be enhanced via a collaborative project. Students should be fairly careful on the following aspects:

(1) Try to understand what the other members exactly mean. Different people may use different words or expressions for the same.
(2) Express what you want to say in a thoughtful way. Harsh words can hurt other members. When a person is emotionally hurt, effective communication becomes almost impossible.

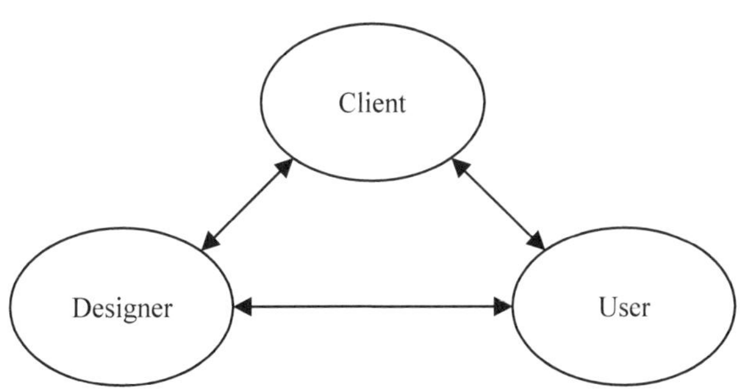

Figure 4.2. Elements of a project

(3) Do not explicitly criticize other members. Try to change your opinion as much as possible. If it is inevitable, do not directly criticize. Instead, do it indirectly.

It should be remembered that a team have the same fate. The ego of each member should be sacrificed as much as possible.

4.4 How to Write the Proposal

At the early stage of a design project, we plan what we will do for the project. We assume there are customers who want something. Then we make a working plan to satisfy the customers. If we write the plan, it is a proposal. In this and subsequent sections, examples of the body of reports are introduced. However, the students do not have to exactly follow the process here. The contents can be varied according to the situation. It is noted that figures and tables help the reader understand the problem. An example of the body of the proposal is as follows:

Title page including team members' names
Summary
The overall process is briefly described.
Statement of Problem
The definition of the problem is described.
Background
The background theory is explained.
Proposed Solution
The plan for solving the problem is described. The flow of the work can be shown by using a figure.
Advantages of Proposed Solution (if needed)
There can be many solution methods. The advantage of the proposed solution is stated.
Disadvantages of Proposed Solution (if needed)
The disadvantage of the proposed method is stated.
Qualification of Personnel and Facilities
Team members can be introduced. The organization of each member can be shown as illustrated in Figure 4.3.
Project Schedule
The project schedule is demonstrated. An example is illustrated in Figure 4.4. It is called a Gantt chart. There are many other ways to show the schedule (Dieter 1999).
Budget
The plan for a budget can be shown by a table as illustrated in Figure 4.5.
References
Quoted references are shown.
Appendix (if needed)

Highly detailed or technical materials are often placed here at the end of the report.

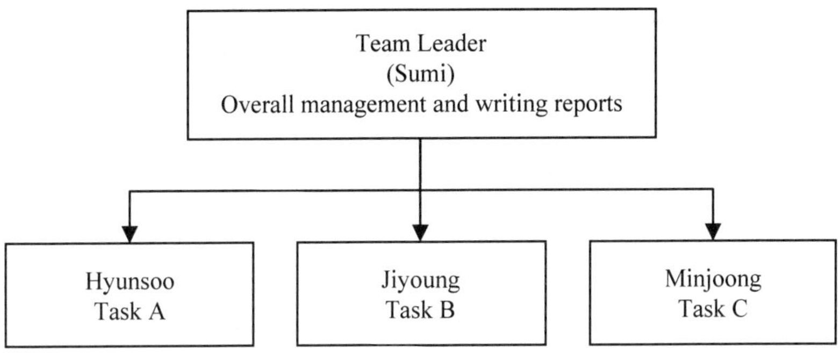

Figure 4.3. Organization of team members

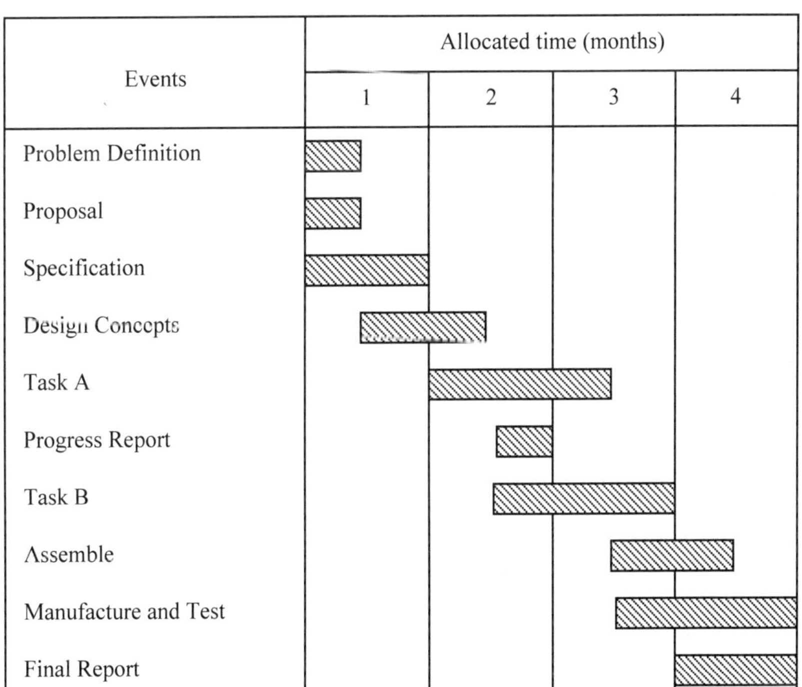

Figure 4.4. An example of a Gantt chart

Item		Funding
Labor		Not applicable to the design project
Direct cost	Equipment	
	Travel	
	Material and supplies	
	Publication and reports	
	Subtotal	
Indirect (overhead) cost		Not applicable to the design project
Total		

Figure 4.5. An example of a budget plan

4.5 How to Write the Progress Report

When the period of a project is long, we may have to write a progress report to examine the progress of the project. The progress report does not have to fully include the description of the project. It should show how the project is processing. The body of the report is as follows:

Title
Description of Problem
Statement of Progress
    The progress is described. Achievement in percentage rate can be shown by comparing the original plan in the Gantt chart.
Amendment of Original Plan
    The original plan can be modified. Modification is explained.
Problems in Progress
    Difficulties of the project are described.
Schedule in Remaining Period
    The Gantt chart can be used.
Summary

4.6 How to Write the Final Report

After the project is finished, a final statement is reported to the client. The body of the final report is as follows:

Title
Abstract
Introduction
  Overall description of the project
Background
Methods or Theories
Working Process (Design process, Experiments)
  How the project has been performed is explained.
Results
  Design results are demonstrated by a drawing, sketch or description of the product.
Conclusions and Future Study
References
Appendix (if needed)

4.7 Final Presentation

To practitioners, most design projects call for a number of meetings and presentations to clients, users, and technical reviewers (Dym and Little 2004). An oral presentation is often given in the meetings and viewgraphs are utilized for visual aid. A designer is the presenter and the presenter should identify the characteristics of the audience. Once the audience has been identified, a team can tailor its presentation to the audience. The presentation must be properly organized and structured.

In the design class, the results of a team project are presented. The audience consists of professors, teaching assistants and peer students. It is fairly important to fully use the given time for the presentation. A longer or shorter presentation is not good. In general, viewgraphs are used for the presentation and a commercial software system called PowerPoint is utilized for the preparation of the viewgraphs (Young and Halvorson, 2004). An example of the viewgraphs is as follows:

Title
  This identifies the title of the project, the client, team members.
Contents
  The order of the presentation is introduced.
Problem statement
  The characteristics of the problem are described.

Objectives
  The key objectives of the design project are presented.
Background
  Background theory is explained.
Organization of the responsibilities
  Organization of the students is presented.
Design process
  Detailed design process is explained. The logic as to why the final design was selected should be mentioned.
Outcome of the design
  The design results are presented.
Design alternatives (if there is)
  If it is possible to make alternatives, show them and explain why they are not selected.
Evaluation of the design results
  The advantages and disadvantages of the final design are explained.
Discussion
  Overall discussion of the final design is made.
Conclusions and future work

The above example is only an example. The viewgraphs can be organized differently according to the situation. The key words should be shown in the viewgraphs because the audience reads while listening. However, the viewgraphs do not have to include details. Generally, having 6-12 lines in a viewgraph is recommended. When a student makes a presentation s/he should not read the viewgraphs word by word. S/he should explain the key words or sentences in the viewgraphs. Figures and brief tables will help the audience understand. Spending just 0.5-1 minute per viewgraph is strongly recommended.

V. References

Antonsson EK, Cagan J (2001) Formal Engineering Design Synthesis. Cambridge University Press, New York

Altshuller G (1996) And Suddenly the Inventor Appeared. Technical Innovation Center, Massachusetts

Beer D, McMurrey D (1997) A Guide to Writing as an Engineer. Wiley, New York

Borowick, JN (2000) Technical Communication and Its Applications. 2nd ed., Prentice Hall, Upper Saddle River, NJ

Cox J, Preppernau J (2007) Microsoft Office PowerPoint 2007 Step by Step, Microsoft Press, Redmond, Washington

Dieter GE (1999) Engineering Design: A Materials and Processing Approach. 3rd ed., McGraw–Hill, New York

Dym CL, Little P (2004) Engineering Design: A Project-Based Introduction. 2nd ed., Wiley, New York

Eggert RJ (2005) Engineering Design. Pearson Prentice Hall, Upper Saddle River, NJ

Haik Y (2003) Engineering Design Process. Brooks/Cole-Thomson Learning, Pacific Grove, CA

Horenstein MN (2006) Design Concepts for Engineers. Pearson Prentice Hall, Upper Saddle River, NJ

Huckin TN, Olsen LA (1991) Technical Writing and Professional Communication. 2nd ed., McGraw-Hill, New York

Hurst K (1999) Engineering Design Principles. Wiley, New York

Hyman BI (1998) Fundamentals of Engineering Design. Prentice Hall, Upper Saddle River, NJ

Lee DG, Suh NP (2006) Axiomatic Design and Fabrication of Composite Structures. Oxford University Press, New York

Young MJ, Halvorson M (2004) Microsoft Office 2003 Inside out, 2003 Edition, Microsoft Press, Redmond, WA

Norton RL (2000) Machine Design: An Integrated Approach. 2nd ed., Prentice Hall, Upper Saddle River, NJ

NSF (1998) Axiomatic Design Workshop for Professors. MIT, Cambridge, MA

Park GJ (2007) Analytic Methods for Design Practice. Springer, Berlin Heidelberg New York

Suh NP (1990) The Principles of Design. Oxford University Press, New York

Suh NP (2001) Axiomatic Design: Advances and Applications. Oxford University Press, New York
Suh NP (2005) Complexity. Oxford University Press, New York
Ullman DG (2003) The Mechanical Design Process. 3rd ed., McGraw-Hill, New York

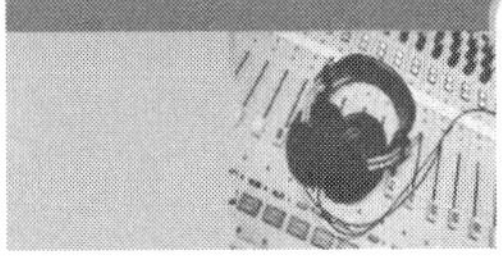

제 1 장 설계기법의 개요

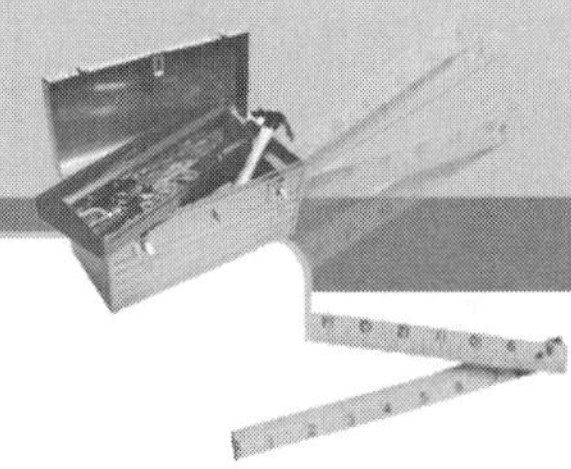

1.1 공학설계란 무엇인가?

설계란 무엇인가? 이에 대한 답은 사람마다 다를 수 있다. 일상생활에 대한 계획을 의미하기도 하고, 예술적 아름다움과 생활의 편리함을 주는 도구 창출을 뜻하기도 한다. 이 책에서 사용하는 설계의 개념은 이런 포괄적인 의미가 아니라 공학적 제품을 생산할 때 세우는 일련의 계획을 뜻한다. 이를 달리 기계설계라고도 한다. 기계설계는 제품의 전반적 기능의 정의나 배치 등일 수도 있고, 자세한 치수의 결정이나 도면 제작을 의미할 수도 있다. 기계설계는 선사시대에 사냥을 위한 도구를 개발할 때부터 존재해 온 무엇보다도 오래된 공학적 행위이다. 그런데 공학적 행위 중 어느 단계를 설계로 보느냐의 문제에 대한 대답은 명확하지 않다. 이렇듯 설계 개념의 불명확성에도 불구하고 일반적으로 공학자들은 계획단계를 설계로 간주하는 데는 대체로 동의하고 있나.

공학적 작업에서 설계의 위치를 살펴보자. 제품의 기획단계에서 판매까지의 흐름도가 그림 1·1에 있다. 그림 1·1에서 음영 처리된 부분이 기계설계 과정이다. 이 과정에서는 기획단계에서 정의한 문제를 공학적 수단을 통해 물리적으로 제조가 가능하도록 변환하는 작업이 이루어진다. 즉, 제품의 기획단계에서 문자나 개념으로 정의된 사항을 물리적으로 구현할 수 있도록 구체화하는 계획이 이루어진다. 설계에서의 주된 작업은 기획단계에서 수립된 사항을 면밀히 검토하고 이의 공학적 타당성을 검증하여 최종 생산에 필요한 각종 부품 및 치수를 결정하는 일이다. 많은 경우 그 결과를 도면화하게 된다. 설계의 최종 산물이라 할 수 있는 도면의 예가 그림 1·2에 있다. 한편, 설계의 목적은 초기의 작업과 후반의 작업을 원활히 연결시키는 데 있다.

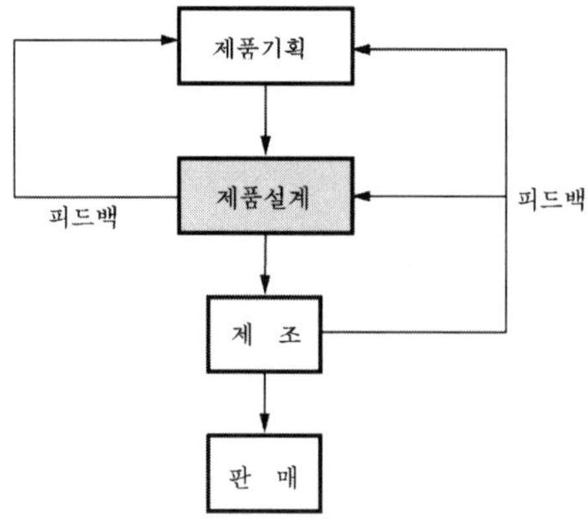

그림 1 · 1 제품생산의 흐름도

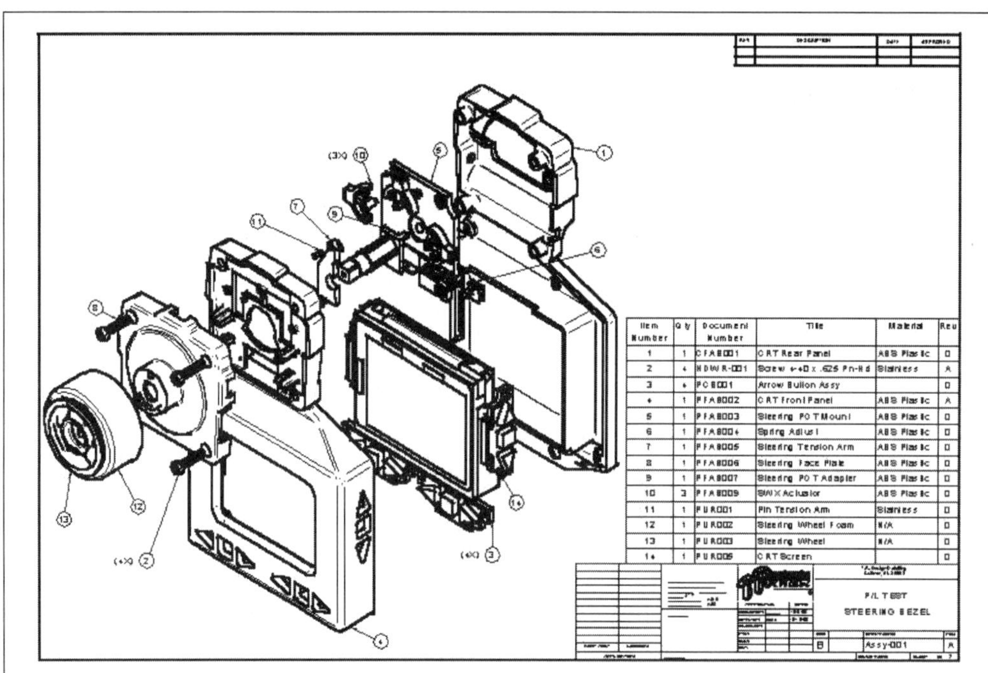

그림 1 · 2 설계도면의 예

설계 과정에서 설계자는 필히 어떤 결정(decision making)을 하게 된다. 그런데 그 결정을 하는 연유나 논리적 과정은 어떤 것일까? 또 좋은 결정의 특성을 객관적으로 정리할 수 있을까? 결정을 내리는 과정에 대한 시각차에 따라 설계를 예술(art)로 보는 쪽과 과학(science)으로 보는 쪽이 있으며 어떤 시각이 옳은지에 대해서는 논란이 계속 되고 있다. 설계를 예술로 간주하는 쪽은 결정 과정에서 직관을 중요하게 여긴다. 설계를 과학으로 간주하는 쪽은 직관보다는 객관적인 논리를 강조하며 보편적인 설계 원리가 존재한다고 주장한다. 예컨대 그림 1·3과 그림 1·4는 자동차의 설계이다. 그림 1·3의 설계에서는 미적인 면이 중요하다. 그러나 그림 1·4에서는 미적인 면보다는 설계조건을 만족하는 기능적인 면이 더 중요하기 때문에 수학적·역학적인 계산을 통해 과학적으로 설계 치수를 결정하게 된다. 이와 같이 설계의 특성은 다양하게 나타나지만 좋은 설계에 대한 보편적인 원리를 찾아낸다면 일반적인 설계법이나 설계이론을 말할 수 있을 것이다. 실제로 최근에는 좋은 설계라고 알려진 것들 사이의 공통점을 찾아내어, 이들을 정형화된 설계이론으로 설명하려는 경향이 나타나고 있다.

그림 1·3 멋진 자동차의 모습

그림 1·4 자동차 내부 구조설계

우리 주변에는 역사적으로 우수한 설계가 많다. 자전거, 자동차, 바퀴, 가전제품 등 우리가 당연하게 받아들이고 있는 제품들 모두가 우수한 설계라 볼 수 있다. 이들 중에는 천재들의 영감에 의해 갑자기 발명된 것도 있고, 노력에 의해 역사적으로 꾸준히 개선되어 온

것도 있다. 지금도 대부분 우수한 특허들은 이런 방식으로 만들어지고 있다. 그런데 설계에서 특정한 이론이나 보편적인 과정을 도출할 수는 없을까? 많은 공학자들 사이에는 과연 설계이론이 존재하는가에 대한 의구심이 팽배하다. 예술적인 면이 고려되는 설계 과정을 정형화된 이론으로 설명하기가 어렵다는 것이 그 의구심의 근거이다. 실제로 예술적으로 훌륭하다고 알려진 제품을 어떤 설계이론으로 설명하는 것은 무리가 있어 보인다. 그러나 사용이 편리하고 성능이 우수한 설계, 즉 공학적으로 좋은 설계에 대해서는 객관적인 설명이 가능할 것으로 보인다. 즉, 공학적으로 좋은 설계를 만들어내는 과정을 객관적으로 분석하고 그들 사이의 공통적인 특징으로부터 좋은 설계에 대한 해답을 얻을 수 있을 것이다. 설계 이론은 이러한 노력으로부터 창출된다.

이 책에서는 이미 확립되어 사용 가능한 설계이론에 대해서만 논의하고자 한다. 예술적이거나 미적인 것을 추구하는 설계에 관한 이론은 그 존재 여부나 객관성 여부의 문제가 있기 때문에 이 책에서는 제외한다. 따라서 공학적 기능을 고려하는 설계에서의 이론적 접근만을 취급한다. 다시 말해 공학적 설계방법론만을 설명한다.

1.2 설계의 특징 및 분류

공학적 작업의 흐름이 그림 1·1에 있다. 제품 기획단계에서는 시장조사 등을 통해 어떤 제품이 필요한가, 그 제품의 기능 및 용도는 무엇인가, 장래의 판매 및 예상이익은 어떤가 등을 결정한다. 이 물음들에 대한 대답을 바탕으로 생산 여부를 결정한다. 기획단계에서 생산을 결정하게 되면 이어서 제품의 구체적인 특성도 함께 결정한다. 이를 두고 광의의 설계라 할 수 있다. 그러나 공학적 설계는 기획이 종료된 후 제품의 구체적인 계획을 수립하는 것을 의미한다. 이미 결정된 제품의 개념을 물리적으로 구현할 수 있는 계획을 확정하는 것이다. 이러한 작업 과정을 설계 프로세스라고 하는데 이 책의 범주를 벗어나므로 참고문헌을 참조하기 바란다.

설계가 전체 공학적 작업에 미치는 영향은 지대하다. 총 생산 비용 중 5%만을 설계 과정에서 사용하나 그 영향은 전체의 80%가 된다는 보고가 있다. 그림 1·1에서 보면 설계 결과가 그 앞의 과정에 미치는 영향은 그리 많지 않다. 그러나 그 후의 제조나 판매에 끼치는 영향은 실로 막대하다. 그림 1·1에서 보듯이 한 단계에서 이전 단계의 오류를 발견하면 피드백 과정에 의해 그 오류의 수정이 이루어진다. 대부분의 경우 피드백 과정은 커다란 손실을 의미한다. 따라서 초기의 작업에서 오류를 최소화하여 손해를 줄일 필요가 있다.

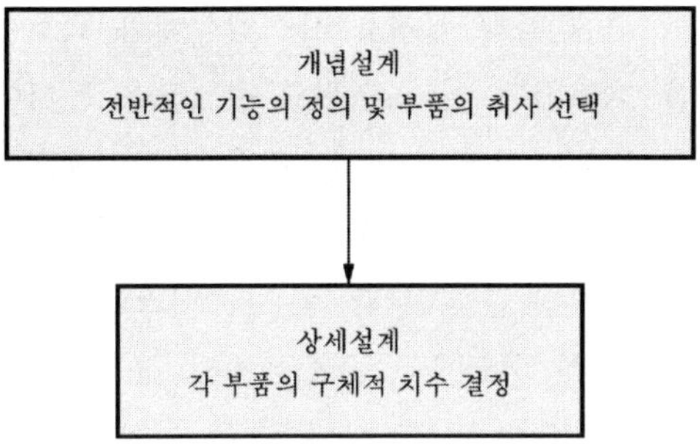

그림 1·5 설계의 순서

설계는 설계방법이나 주위의 환경에 따라 다르게 분류할 수 있다. 우선 과정의 선후에 따라 그림 **1·1**의 음영 처리된 부분을 개념설계와 상세설계로 나눈다. 설계의 순서는 개념설계, 초기설계, 상세설계의 세 가지로 나누기도 한다. 개념설계는 대상제품의 선정 및 기능의 정의, 초기설계는 각 부품의 선정 및 배치 등을 하는 것이다. 개념설계와 초기설계에는 비슷한 설계방법을 사용한다. 따라서 설계방법의 적용면에서 설계단계를 개념설계와 초기설계를 합한 개념설계와 상세설계로 분류한다. 두 가지의 순서도가 그림 **1·5**에 있다. 개념설계에서는 영향이 큰 사항이 대부분 완성된다. 반면 상세설계에서는 개념설계 단계에서 결정한 설계 결과를 바탕으로 각 부품의 치수, 모양 등을 결정한다. 이때 제품의 성능을 판별하기 위한 공학적인 실험이나 해석이 수반된다. 이러한 상세설계의 전형적 과정이 그림 **1·6**에 있다.

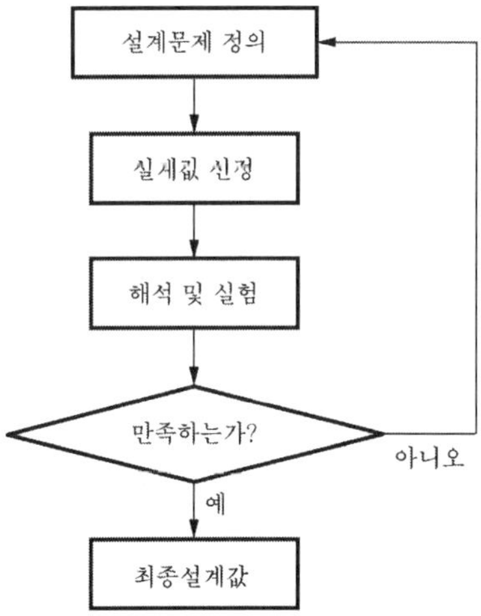

그림 1·6 상세설계 과정

한편으로는 결정할 값의 수에 따라 설계를 시스템설계와 요소설계로 나누기도 한다. 시스템설계는 주위의 환경을 고려하여 여러 부품의 설계값을 동시에 구하는 경우를 말한다. 시스템설계에서는 부품 사이의 상관관계나 상호작용을 고려하여 설계조건을 만족하는 시스템을 결정한다. 이와 반대로 요소설계는 한가지 부품의 설계만을 결정한다. 선반을 설계하는 경우를 예로 들면, 기계 전체를 설계할 경우가 시스템설계이며 주축 하나만 설계하는 경우가 요소설계이다. 그러나 최근에는 요소설계의 경우에도 주위의 영향을 고려해야 할 때 시스템설계와 같은 방법을 취하기도 한다. 축을 설계할 때 주변의 베어링이나 키 등의 영향을 고려하는 것이 그러한 경우에 해당한다.

시작단계에서의 의도에 따라 설계를 창의적 설계, 모방설계, 개선설계 등으로 분류하기도 한다. 창의적 설계는 존재하지 않는 제품을 창출하려는 노력이다. 주로 기술적으로 앞선 곳에서 채택되는 창의적 설계는 새로운 기능을 발명한다. 창의적 설계는 성공이 매우 어려운 만큼 흔히 수행되지는 않으나 다른 분야에 미치는 영향이 대단히 큰 것이 특징이다. 전통산업에서는 드문 반면, 전자산업에서는 많이 보인다. 최근의 휴대폰, DVD, MP3 등이 창의적 설계에 의해 출현한 것으로 볼 수 있다. 모방설계는 대부분 후발 주자들이 사용하는데, 이미 존재하는 제품의 특허를 피해가면서 기존 제품보다 성능이 우수한 제품을 계획하는 작업이다. 이 작업은 비교적 용이하나 다른 분야에 미치는 영향이 크지 않다. 개선설계는 가장 많이 사용하는 방식으로 기존의 제품에 기능 변경을 시도하거나 새로운 기능을 추가할 때 사용한다. 이미 창의적 설계에 의한 제품을 보유하고 있을 경우, 그 제품 설계에 관한 노하우(know-how)가 있으므로 개선을 통해 다양한 제품을 설계할 수 있다.

최근에는 앞에서 분류한 여러 종류의 설계에 객관적이고 과학적인 설계방법을 적용하려는 노력이 일고 있다. 다시 말하면 설계 과정에 과학적이고 정형화된 방법을 사용하려는 시도가 이루어지고 있다. 모든 설계분야에 항상 적용이 가능한 설계법이 존재한다면 매우 다행스러운 일이겠지만, 지금으로서는 그러한 방법이 존재하지는 않는 것으로 보인다. 따라서 설계 과정의 전후나 설계환경의 특성에 따라 적당한 방법을 적용하는 것이 현명하리라 생각한다. 물론 각 방법 간에 공통적인 양상이 있기는 하지만 설계의 특징, 환경, 목적에 따라 적절한 방법을 선택하는 것이 일반적이다.

1.3 설계방법론

설계방법론이란 모든 설계행위에서 공통적으로 사용할 수 있는 일반적이고도 과학적인 이론이자 원리라 할 수 있다. 과학적이라 함은 객관적, 논리적 타당성을 가지는 것을 뜻한다. 예술이 아닌 공학적 설계에서조차 설계방법론이 존재하는지 여부의 문제는 설계 연구자들의 최대 관심사이다. 설계방법론의 존재를 의심하는 이들은 설계의 직관적인 면을 강조한다. 실제로 개념설계 단계에서는 직관이나 경험에 의한 작업이 주를 이루는 경우가 많다. 반면에 개념설계를 완성한 후, 상세설계단계에서는 대상 제품의 특성을 평가하기 위하여 그림 1·6과 같이 과학적인 분석이나 해석방법을 동원한다. 이 경우 과학적인 해석 결과를 관련 이론에 입각하여 설계에 반영하는데, 이 과정에서조차 숙련자의 경험이나 직관을 이용하는 경우가 많다. 따라서 객관적으로 계산된 해석 결과를 설계에 반영할 때, 그 반영 정도가 설계자 개인의 자의적 판단에 좌우되게 된다. 이 때문에 그림 1·6의 방법적 효율성과 적합성에 대해 회의적인 시각이 존재한다. 이에 따라 최근에는 좋은 설계에 공통적으로 존재하는 원리를 찾아서 응용 가능한 설계방법론을 찾으려는 노력이 일고 있다.

설계방법론은 좋은 설계 사례들에 공통적으로 존재하는 원리를 규명함으로써 만들어진다. 대표적인 것이 공리적설계이다. 공리적설계는 모든 설계에 공히 적용될 수 있는 공리로 이루어져 있다. 좋은 설계를 위해 공리적설계에서는 두 가지 공리를 주장한다. 기능요구를 독립적으로 만족시키는 설계파라미터를 정해야 한다는 독립공리와, 관여된 정보량을 최소화해야 한다는 정보공리가 그것들이다. 최근 많은 예제들이 발표되어 공리적설계가 유용함을 보였다. 한편 설계 시에 사고 과정을 도와주는 도구로 TRIZ라는 것이 있다. TRIZ에서는 TRIZ가 요구하는 방법대로 설계 과정을 거치면서 분석하고 도표를 만든다. 이때 분석한 결과와 데이터베이스에 저장되어 있는 우수한 특허 사례를 비교함으로써 새로운 설계를 도출하는 데 도움을 받을 수 있다. 그 외에도 각 설계의 예마다 프로세스를 정의하는 방법이 있다. 설계의 순서를 논리적으로 정하여 효율화를 꾀한다. 이상에서 언급한 설계법은 일반적으로 개념설계 과정에서 많이 사용한다.

설계방법론을 가장 활발하게 사용하는 곳으로 상세설계를 들 수 있다. 상세설계 시에 지난 수십 년 동안 많이 사용된 도구는 최적설계기법이다. 최적설계에서는 목적함수와 설계조건으로 이루어진 제한조건을 수학적으로 정의한 후, 그 설계조건을 만족하면서 목적함수를 최소화하는 설계변수를 결정한다. 최근에는 컴퓨터 성능과 해석기법의 발달로 인하여 대규모 문제를 용이하게 풀 수 있게 되어, 완벽한 설계방법론으로는 최적설계가 거의 유일하다고 할 수 있다. 그러나 모든 문제를 수학적으로 표현해야 하기 때문에 실제 공학적 가치관을 최적설계문제에 반영하기 어려운 경우가 많다. 최적설계의 대안 혹은 보조 도구로서 문제의 특성에 따라 실험계획법, 강건설계법 등 여러 방법이 제시되고 있으나, 문제를 바라보는 관점이 기본적으로 유사할 뿐만 아니라, 각 방법들 자체의 여러 문제점이 여전히 남아 있다. 한편 상세설계는 개념설계의 개념이 완성된 후 자세한 치수 등의 결정에 이용하기 때문에 제품에 미치는 영향력이 개념설계에 비해서는 그렇게 크지 않다.

개념설계 및 상세설계 이외에 앞에서 분류한 다른 설계분야에도 그 특성에 따라 위에서 언급한 설계방법을 이용할 수 있다. 지금부터 이 책에서 다루는 각종 설계방법에 대해 간단히 논하기로 한다.

1.4 공리적설계

공리적설계에서는 설계를 '무엇을 이룰 것인가?'와 '그것을 어떻게 얻을 것인가?'의 상호작용이라고 정의한다. 이때 설계자는 양쪽의 기능 및 능력을 적절히 조합하고 배치해서 원하는 것을 얻으려고 한다. 설계를 수행할 때 문제를 정의하고 설계를 완성하는 등의 순서에 따라 그림 1·7과 같이 영역(domain)을 나눌 수 있다. 소비자 영역에서는 소비자들이 원하는 요구를 정의하는데, 대개 이들의 요구는 서술적으로 표현할 수 있다. 그런데 최근에는 소비자가 필요로 할 것 같은 사항을 생산자가 미리 예측하여 정의하는 경우도 흔하다. 소비자 영역에서 이렇게 정의한 것을 소비자요구(customer attributes, CAs)라 부른다. 이렇게 소비자 영역에서 정의한 사항을 기능적 영역에서 기능요구(functional requirements, FRs)로 변환한다. 기능요구는 소비자 영역의 소비자요구를 공학 용어로 다시 기술한 것이다. 이는 앞서 언급한 '무엇을 이룰 것인가?'에 대한 공학적 정의라 할 수 있다. 이러한 기능요구를 만족시키려면 물리적 영역에서 설계파라미터(design parameters, DPs)를 정의해야 한다. 대개 이 과정을 설계 과정이라 부른다. 설계파라미터를 결정한 후에는 동일한 방법으로 설계파라미터로부터 제조를 위한 생산변수(production variables, PVs)를 정의하게 된다. 이와 같이 공리적설계에서는 이웃하는 두 영역 사이의 관계를 파악하여, 앞의 영역에서 정의한 사항에 따라 뒤따르는 영역의 사항을 정의하게 된다. 이를 영역간 사상이라 부른다. 좋은 설계란 이 사상을 우수하게 정의한 상태를 뜻하기도 한다.

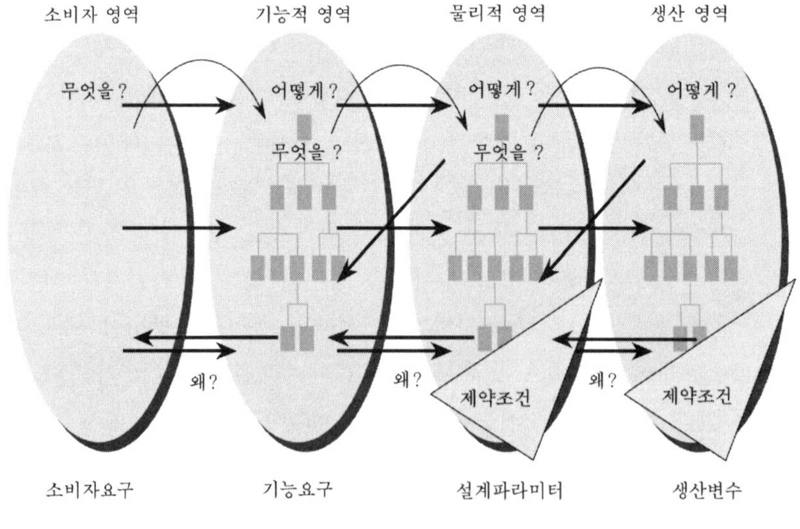

그림 1·7 영역, 사상, 설계공간의 관계

설계공리는 모든 공학 행위의 근간을 이루는 공통적인 원리를 집약적으로 표현한 것으로 다음과 같이 정의한다.

제 1 공리 : 독립공리 (The Independence Axiom)
기능요구의 독립성을 유지하라 (Maintain the independence of the functional requirements(FRs)).

제 2 공리 : 정보공리 (The Information Axiom)
설계 대상의 정보량을 최소화하라 (Minimize the information content of the design).

위의 공리는 간단하게 보이지만 공학적으로 많은 의미를 함축하고 있다. 제 1 공리는 일반적으로 설계자들이 의식적이건 무의식적이건 알고 있는 직감의 표현이라 할 수 있다. 독립공리의 의미는 복잡한 시스템을 설계할 경우 적절한 설계파라미터(DPs)를 설정하여 주어진 기능요구(FRs)들을 독립적으로 만족시키는 것이 좋은 설계라는 것이다. 그림 **1 · 7**의 기능적 영역에서 정의한 기능요구를 물리적 영역에서 정의한 설계파라미터를 사용하여 각각 독립적으로 만족시켜야 한다는 것을 말하고 있다. 다시 말해 각각의 기능요구들을 나누어서 정복하라는 의미이다. 제 2 공리는 제 1 공리를 만족할 수 있는 복수의 설계를 발견했을 경우, 그중 성공의 확률로 정의되는 정보량을 최소화하는 설계를 선택하라는 의미이다. 이것도 역시 설계자들이 평소에 직관적으로 느끼고 있는 것과 일치한다. 뒤에 설명하겠지만 제 2 공리는 강건설계(robust design)와도 연관이 있다. 이렇게 공리는 비교적 간단하게 설명되지만 직접 설계에 적용하는 것은 상당히 어려울 때도 있다. 공리들을 적용하는 감각이나 그때의 독립성이나 정보량을 정의하고 느끼는 데는 많은 시간을 요할 수도 있다.

공리적설계는 대개 다음과 같은 설계분야에서 응용할 수 있다.

(1) 개념설계단계에서의 창의적 설계
(2) 기존에 있는 복수의 설계에 대한 비교분석
(3) 현재의 설계분석으로부터 개선설계안의 도출

이에 대한 자세한 사항은 뒤에서 설명하도록 한다. 한편 공리는 대개 기하학에서 정의해 왔는데, 기하학의 경우와 마찬가지로 여기서도 공리로부터 증명이 가능한 많은 정리 및 추론을 도출할 수 있다.

1.5 최적설계

최적화이론은 수학적으로 다음과 같이 정식화된 문제의 해를 구하는 방법에 관한 이론이다.

$$\text{Find} \quad \mathrm{b} \in R^n \tag{1·1a}$$

$$\text{to minimize} \quad f(\mathrm{b}) \tag{1·1b}$$

$$\text{subject to} \quad h_i(\mathrm{b}) = 0, \quad i = 1, \cdots, l \tag{1·1c}$$

$$g_j(\mathrm{b}) \le 0, \quad j = 1, \cdots, m \tag{1·1d}$$

$$\mathrm{b}_\mathrm{L} \le \mathrm{b} \le \mathrm{b}_\mathrm{U} \tag{1·1e}$$

여기서 b는 n개의 요소로 이루어진 설계변수벡터, f는 목적함수, h_i는 i번째 등제한조건, g_j는 j번째 부등제한조건, b_L과 b_U는 각각 설계변수의 하한, 상한벡터이며, l은 등제한조건의 수, m은 부등제한조건의 수이다.

최적화이론에서는 주어진 목적함수를 최소화하거나 최대화한다. 이 점이 공학설계에서 특정 공학적 특성을 극대화하려는 것과 일치하는 면이다. 공학설계에 최적화이론을 도입할 경우 이를 최적설계라 부른다. 최적설계에서는 식 (1·1b)의 목적함수가 그 설계에서 추구하는 공학적 특성을 나타내며, 각종 주변 조건들이 식 (1·1c)와 (1·1d)의 제한조건들이 된다. 최적화이론을 공학설계에 도입할 때는 그 설계문제를 식 (1·1)과 같은 형태로 정식화할 수 있어야 한다. 최적설계는 설계변수의 값을 구체적으로 결정할 수 있는 점 때문에 주로 상세설계에서 사용된다. 개념설계 과정에서는 부품의 선택이나 배치 등을 결정하게 되므로 개념설계를 식 (1·1)과 같이 표현하기는 어렵다. 따라서 전반적인 설계 흐름상에서 보면 개념설계단계에서 설계 대상의 대부분이 개념적으로 잘 정의된 상태에서 구체적인 치수를 결정하는 데 최적설계가 사용된다고 보는 것이 타당하다. 최적설계를 적용할 때는 식 (1·1)과 같이 정식화한 문제의 해를 구하도록 특별히 개발된 알고리즘을 사용한다.

최적설계는 정식화를 의미 있게 정의한 경우에 우수한 설계해를 제공한다. 그러나 정식화가 현상을 분명하게 표현하지 못하거나 정식화의 공학적 의미가 실제 상황과 다를 경우에는 좋지 않은 설계해를 제공할 수도 있다. 한편 최적설계에서 얻은 해는 일반적으로 제한조건을 극한적으로 만족한다. 이 경우 설계변수에 공차가 있거나 약간의 변화가 있다면 그 해는 제한조건을 만족하지 못할 수 있다. 즉, 제한조건의 만족여부가 매우 중요한 문제에는 최적설계를 적용하는 데 다소 어려움이 있다. 물론 이를 극복하기 위해 여러 기술들이 제안되고 있다.

최적설계는 그 이름이 공학적으로 추구하는 가치와 잘 일치한다는 점에서 우리로 하여금 마치 모든 설계문제를 단숨에 해결할 것 같은 희망을 가지게 할 때가 있다. 이는 최적화이론을 설계에 응용할 때에 특히 경계해야 할 점이다. 최적설계는 설계문제를 식 (1・1)의 형태로 표현할 수 있을 때에만 적용 가능하다는 점을 유념해야 한다. 특히 부품의 선택 등 전반적인 계획이 완료된 후에 상세설계 과정에 최적설계를 응용하는 것이 효율적이다. 또한 개념설계에 비해 설계 과정에서의 영향도가 적다는 점도 기억해야 한다. 따라서 최적설계를 응용할 때에는 주변 사정을 잘 고려하여 주의를 기울여야 한다.

1.6 구조최적설계

구조물의 설계는 최적화이론을 가장 많이 응용하는 분야 중의 하나이다. 대상 구조물의 구체적 모양이나 치수를 결정할 때에 최적설계기술을 사용할 수 있다. 최근에는 유한요소법(finite element method)의 발전과 더불어 최적설계의 응용분야인 구조최적설계기술이 확산되고 있다.

여기서는 구조최적설계에서 가장 많이 사용하는 선형정역학 지배방정식을 살펴본다. 지배방정식을 유한요소법으로 표현하면 일반적으로 다음과 같다.

$$K(b)z = f \qquad (1 \cdot 2)$$

여기서 K는 강성행렬, z는 절점에서의 변위벡터, b는 치수 등으로 이루어진 설계변수 벡터, f는 외력벡터이다.

실제문제에서 최적설계를 사용할 때에는 다루고 있는 시스템의 지배방정식을 컴퓨터로 계산해서 목적함수나 제한조건을 계산하게 된다. 지배방정식은 식 (1・1c)의 등제한조건으로 취급한다. 실제 최적화 과정에서는 식 (1・2)의 등제한조건을 풀어서 식 (1・1d)에 대입하여 등제한조건을 소거한다. 일반적인 구조최적설계의 문제는 다음과 같이 정식화할 수 있다.

$$\text{Find} \quad b \in R^n \qquad (1 \cdot 3a)$$

$$\text{to minimize} \quad f(b, z) \qquad (1 \cdot 3b)$$

$$\text{subject to} \quad K(b)z = f \qquad (1 \cdot 3c)$$

$$g_j(b, z) \le 0, \quad j = 1, \cdots, m \qquad (1 \cdot 3d)$$

$$b_L \le b \le b_U \qquad (1 \cdot 3e)$$

여기서 목적함수를 나타내는 식 (1・3b)는 일반적으로 구조물의 중량으로 정의하지만 각종 반응값을 사용할 수도 있다. 부등제한조건인 식 (1・3d)는 변위, 응력, 고유진동수 등에 대한 한계값을 의미한다. 최적화 알고리즘을 식 (1・3)에 적용할 때에는 식 (1・3)에 있는 함수들의 설계변수에 대한 민감도가 필요하다. 민감도란 미분값을 의미하는데 이를 구하는 과정은 어렵고 비용이 많이 든다. 따라서 구조최적설계에서 중요한 문제는 민감도를 효율적으로 계산하고 이용하는 것이다.

구조최적설계는 설계변수의 특징에 따라 추구하는 바나 사용방법이 달라진다. 구조최적설계는 설계변수의 성격에 따라 표 **1・1**과 같이 분류할 수 있다. 구조물을 설계할 때에는 구조물이 보이는 다양한 물리적 현상을 규명하기 위해 다양한 해석방법을 동원한다. 이러한 해석들을 최적화 과정에서 사용하려면 각종 해석에 대한 지배방정식은 식 (1・3c)의 등제한조건으로 취급한다. 그런데 해석이 복잡할 경우에는 앞서 언급한 민감도해석이 복잡해진다. 민감도에 관한 많은 연구가 있었지만, 민감도를 해석적으로 구하는 과정이 수치적으로 일관되게 구현되어 컴퓨터에서 사용 가능한 경우는 다루는 해석이 선형정역학일 경우뿐이다. 따라서 진동/소음 해석, 시간영역에서의 동역학해석, 비선형해석, 열전달 해석, 유동해석 등을 이용하는 경우에는 수학적 최적화가 아직까지 다소 제한적이다. 예를 들어 시간영역에서 제한조건 등을 정의할 경우에는 제한조건의 수가 매우 많기 때문에 이를 용이하게 처리하기란 쉽지 않다.

표 1・1 구조최적설계의 분류

구조최적설계의 종류	설계변수의 특징
치수최적설계	단면계수, 두께 등이 설계변수이다. 설계변수의 변화 시 유한요소법의 영역은 고정된다.
형상최적설계	설계변수가 구조물의 형상을 표현한다. 설계변수의 변화 시 유한요소법의 영역도 함께 변화한다.
배치최적설계	유한요소법의 변환행렬이 설계변수의 함수이다. 자주 사용하지 않는 방법이며 형상최적설계와 동일하게 분류하기도 한다.
위상최적설계	유한요소법의 요소의 존재 여부가 설계변수이다.

1.7 동하중이 가해지는 구조물의 최적화

동하중이 가해지는 구조물은 시간에 따라 반응의 크기와 방향이 변한다. 따라서 동하중이 가해지는 구조물을 최적화하는 경우에는 구조물에 나타나는 반응의 시간종속성을 고려해야 한다. 구조물에 나타나는 반응의 시간종속성을 고려하여 최적화를 시도할 경우를 구조물의 동적반응최적화라고 한다. 구조물에 가해지는 힘은 실세계에서는 대부분 동하중이다. 실세계의 구조물을 최적화한다는 것은 엄밀한 의미에서 구조물의 동적반응을 최적화하는 것이다. 그러나 실제문제에 사용되는 구조최적화 기술은 앞서 소개한 정적반응최적화기법이 대부분이다. 이는 동적반응최적화가 정적반응최적화에 비해 몇 가지 어려운 점을 가지고 있어서 그 연구가 정적반응최적화 연구에 비해 활발하게 이루어지지 않았기 때문이다. 여기서는 구조물의 동적반응최적화가 무엇이며, 동적반응최적화를 수행하기 위해 필요한 요소들은 무엇인지에 대해 대략적으로 기술한다.

시간을 변수 t로 할 때 동하중 $\mathrm{f}(t)$가 구조물에 가해질 경우의 평형방정식은 다음과 같다.

$$\mathrm{M}(\mathrm{b})\ddot{\mathrm{z}} + \mathrm{K}(\mathrm{b})\mathrm{z} = \mathrm{f}(t) \tag{1・4}$$

여기서 M은 질량행렬, $\ddot{\mathrm{z}}$는 가속도벡터이다. 한편 식 (1・5)는 구조물의 동적반응최적화를 위한 정식화를 일반적인 형태로 표현한 것이다.

$$\text{Find} \quad \mathrm{b} \in R^n \tag{1・5a}$$

$$\text{to minimize} \quad f(\mathrm{b}, \mathrm{z}, t) \tag{1・5b}$$

$$\text{subject to} \quad \mathrm{M}(\mathrm{b})\ddot{\mathrm{z}} + \mathrm{K}(\mathrm{b})\mathrm{z} = \mathrm{f}(t) \tag{1・5c}$$

$$g_j(\mathrm{b}, \mathrm{z}, t) \le 0, \quad j = 1, \cdots, m \tag{1・5d}$$

$$\mathrm{b}_\mathrm{L} \le \mathrm{b} \le \mathrm{b}_\mathrm{U} \tag{1・5e}$$

식 (1・4)는 앞서 설명한 바와 같이 등제한조건식 (1・5c)로 사용된다. 식 (1・5)의 설계해를 구하는 것이 구조물의 동적반응최적화이다. 앞 절의 구조최적설계 정식과 비교하면 단지 등제한조건의 지배방정식만이 바뀌고 제한소선이 시간의 함수이다.

이 정식화의 특징은 제한조건이 시간의 함수라는 것이다. 위의 정식화를 풀기 위해 많은 경우 컴퓨터를 이용한 수치적 방법을 택하는데 수치적 방법에서는 시간을 이산적으로 나누어야 한다. 따라서 시간종속함수가 포함된 식 (1・5c)나 (1・5d)의 식은 이산적으로 나누어진 시간의 수만큼 생성된다. 일반적으로 이 수가 매우 많은 편이므로, 많은 수의 제한조건을 어떻게 효율적으로 처리하느냐가 동적반응최적화의 관건 중 하나이다. 때문에 시간종속함수를 처리하는 여러 방법들이 제시되어 왔다. 시간종속함수를 적분하여 시간에 종속적이지 않은 범함수로 변환하거나, 전체 시산영역에서 함수가 극값(peak value)을 가지는 경우만을 고려하는 방법 등이 그 예이다. 이들은 고려하려는 문제에 따라 성공적으로 적용될 수도, 그렇지 않을 수도 있다. 그 까닭은 범함수나 민감도를 계산하는 과정이 복잡하고 비용이 많이 들기 때문이다. 또한 극값을 고려할 때는 극값의 수가 많으므로 민감도계산이 어렵고 극값이 설계조건에 따라 변하는 등의 문제점도 있다. 이런 이유로 시간종속함수의 처리 기법은 규모가 작은 문제에 적용된 예는 있으나 대형의 문제에 적용된 예는 거의 없다.

한편 최근에 동적평형방정식을 정적으로 변환하여 반복 과정을 통해 동적반응최적화를 해결하려는 경향이 있다. 정적반응최적화는 이미 잘 개발되어 있다. 이 방법은 동하중을 시간영역에서 이산화하여 다수의 정하중으로 변환한 후, 변환된 다수의 정하중을 다중하중조건으로 처리하여 정적반응최적화를 반복적으로 시행한다.

1.8 실험계획법에 의한 설계

시스템의 특성이나 성질을 규명하기 위해 각종 실험을 할 때가 있다. 실험 후에는 측정하거나 산출한 결과를 목적에 맞게 분석한다. 분석할 때에는 특성치에 영향을 주는 수많은 요인들을 찾아내고 찾아낸 요인의 값을 결정하여 원하는 실험 결과를 얻는다. 이렇게 조직적으로 실험을 하고 결과를 분석하는 방법으로 실험계획법(design of experiments)이 있다. 최근 들어 실험계획법에서 사용하는 방법을 설계에 도입하려는 경향이 있다.

실험계획법을 이용하는 설계법은 실험계획법으로 대상 제품의 설계 모형을 수립하고 수립된 모형을 이용하여 설계값을 찾는다. 이때 실험 결과는 실제 실험을 통해 얻을 수도 있지만 컴퓨터로 계산한 결과를 사용할 수도 있다. 실험계획법에 의한 설계는 일반적으로 모형의 모형이라 할 수 있는 메타모형(meta model)을 이용하고 있다. 즉, 주어진 설계점에서 얻은 실험값이나 함수 계산값을 사용하여 원래의 실험 양상이나 함수를 근사화하는 근사모형(근사함수)을 만든 후, 만든 근사모형으로 최적설계 등을 실시한다. 메타모형을 사용하면 실험 횟수나 계산량을 절약할 수 있는데 크게 두 가지 방법으로 설계에 접근한다.

첫 째, 실험계획법에서 바람직한 실험 결과를 얻기 위해 요인의 값을 결정하는 특정 방법을 그대로 사용하는 경우이다. 실제 실험에서의 각종 요인을 설계변수로 하고, 실제 실험에서 얻는 결과값을 목적함수로 하여 설계를 수행한다. 실험이나 계산이 고가가 아닐 경우에는 가능한 모든 변수 조합에 대해 실험 결과를 계산하는 전조합실시법(factorial design)을 수행할 수 있다. 그러나 실험이나 계산이 고가일 때는 일부실시법을 실시한다. 대표적인 일부실시법으로 직교배열표(orthogonal array)를 이용하는 방법이 있다. 직교배열표는 특히 이산공간(discrete space)에서 치수를 구할 수 있는 등 일반적인 상세설계의 특성과 일치하는 점이 많아 사용하기 쉽다. 그러나 직교배열표를 효율적으로 응용하기 위해 많은 주의가 필요하다.

둘 째, 반응표면법(response surface method)을 사용하는 경우이다. 반응표면법은 메타모델을 이용하는 대표적인 방법이다. 이 방법은 주어진 여러 개의 설계후보점에서 실험이나 계산값을 산출한 후 설계후보점과 산출한 값의 관계를 명시함수(explicit function)로 근사화한다. 그 명시함수의 형태는 대부분 2차 함수이다. 이때 최소자승법(least square method)을 사용하여 실제 실험값이나 함수 계산값과 근사화된 함수값과의 차이를 최소화한다. 일단 근사화된 함수가 결정되면 최적설계방법 등을 사용하여 설계값을 구할 수 있다. 이때 설계후보점을 선정하는 방법에 따라 근사화의 정확도가 달라질 수 있으므로 설계후보점을 선정하는 데 주의를 기울여야 한다.

실험계획법에 의한 설계에서는 설계변수의 수가 많으면 설계를 진행하기 어려운 것으로 알려져 있다. 또한 의미 있는 결과를 얻기 위해서 여러 가지 통계적인 이론으로 설계 결과를 검증해야 한다. 그러나 이런 단점에도 불구하고 설계이론에 관한 해박한 지식이 없이 쉽게 사용할 수 있다는 장점 때문에 실험계획법을 사용하는 경우가 점점 증가하고 있다.

1.9 강건설계

제품 성능뿐 아니라 제품 성능의 질도 향상시키려는 노력을 제품설계단계에서 하는데, 그 중의 하나가 강건설계(robust design)이다. 강건설계가 추구하는 바는 설계변수의 공차(tolerance)나 정의하기 어려운 주변의 변화(noise 혹은 perturbation)에 대해 제품성능의 변화를 최소화하여 제품의 품질을 유지하고자 하는 것이다. 일본의 다구치(Taguchi)가 강건설계 개념을 실험계획법에 적용하여 공차나 주변 변화에 둔감(insensitive)한 인자를 찾는 방법을 제안한 이래 강건설계법은 기계설계에 계속 적용되고 발전되어 왔다.

설계를 마치면 설계변수값을 정하게 되는데, 이를 공칭설계(nominal design)라 한다. 공칭설계에 의해 계산되는 제품성능(최적화의 용어로는 목적함수로 볼 수 있다)을 공칭값(nominal value)이라 부른다. 그러니 흔히 설계변수에는 공차(변동)가 있다. 한편 주변의 온도나 습도 등 설계문제에서 감안하기 어려운 변수들에도 변동이 있다. 이러한 변동을 잡음(noise)이라 부르기도 한다. 변동이 존재하면 공칭값에도 변동이 발생하기 마련이다. 주변의 조그만 변동에 대해 공칭값의 변화가 큰 설계를 변동에 대해 민감(sensitive)한 설계라고 말하고, 일반적으로 좋지 않은 설계로 간주한다. 주변의 변동에 대해 성능이 목표치 주위에 좁게 분포하는 설계를 변동에 대해 둔감(insensitive)한 설계라 하는데, 변동에 대해 둔감한 시스템을 설계하는 기법이 강건설계이다.

강건설계의 구현에는 몇 가지 방법이 있는데 이들을 설명하면 다음과 같다. 먼저 다구치 방법을 그대로 이용하는 경우가 있다. 이는 실험이나 계산 횟수를 줄이기 위해 직교배열표에서 공차를 감안하여 반응치를 구한 후, 반응치의 분포를 최소화하는 방법이다. 이때 반응치의 분포를 표현하는 방법으로 손실함수(loss function)나 S/N(signal-to-noise)비율을 사용하기도 한다. 다구치의 제안 이래 이 방법의 문제점이 여러 번 지적되었으나 아직도 많이 사용하고 있다. 강건설계를 구현하는 다른 방법은 최적설계에서 목적함수를 대상함수의 민감도로 대치하여 최적화를 수행하는 것이다. 이 과정에서는 민감도의 민감도, 즉 2차 미분값이 필요하므로 강건최적설계는 고가의 방법이 될 수 있다. 이를 극복하기 위해 변형된 기술들을 사용하기도 한다. 마지막으로 공리적설계방법을 강건설계에 응용할 수도 있다. 강건성은 공리적설계에서 성공의 확률로 표현하는 정보량과 관계 있는 것이 확인되었다. 따라서 제 1 공리를 적용한 후, 복수의 설계 중 하나를 선정하는 과정에서 강건설계를 얻을 수 있다. 한편 기능요구가 하나일 때는 설계강성의 크기를 작게 하면 강건설계를 얻을 수 있다. 기능요구가 하나 이상일 때에도 설계행렬의 대각항들의 크기를 작게 하는 것이 강건설계의 개념과 일치한다.

1.10 다분야통합최적설계

지금까지 상세설계를 할 수 있는 다양한 설계기법을 소개하였다. 이런 다양한 상세설계 기법들은 모두 최적설계라는 큰 범주에 속한다. 최적설계는 광범위하게 적용되고 있어 그 적용 사례를 많이 발견할 수 있다. 그런데 최적설계를 적용한 사례들의 일반적인 특징 중의 하나는 시스템의 특성을 알기 위한 해석이 하나의 분야에서만 이루어진다는 점이다. 그러나 실제로는 여러 분야를 고려하여 대상 시스템의 문제를 해결하는 수가 많다. 여러 분야를 동시에 고려하는 실제 공학문제를 해결하기 위해 다양한 분야를 고려하는 설계방법론이 대두되었는데, 이것이 다분야통합최적설계(multidisciplinary design optimization, MDO)이다.

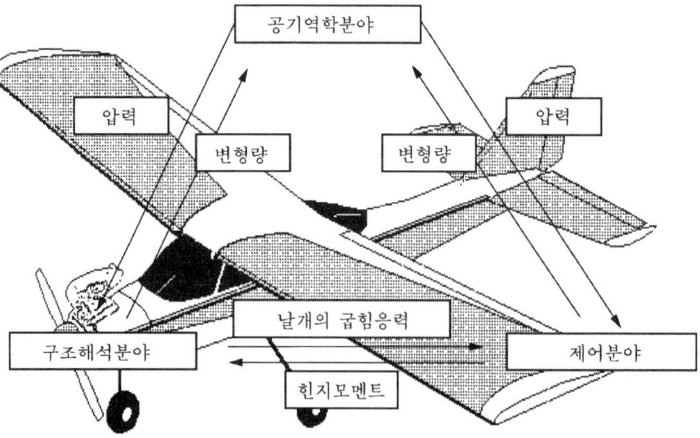

그림 1·8 전형적인 해석간 연성관계

다분야통합최적설계를 이해하기 위해서는 우선 고려하고 있는 해석들 사이의 관계를 살펴보아야 한다. 다분야통합최적설계의 주된 관심은 해석들 사이의 연성이다. 여기서 연성이란 한 분야의 해석 결과가 다른 분야들의 해석에 입력으로 작용하고 그 역도 성립하는 관계를 말한다. 이 연성 때문에 해석들 사이에는 복잡한 관계가 생겨 전체적인 해석이 어려워진다. 그림 **1·8**에 전형적인 연성관계가 있는데 이를 최적화 과정에서 수용하기란 쉽지 않다.

최적설계를 수행할 때에 연성관계가 있는 해석이 필요할 때, 구체적인 최적화방법은 문제마다 다양하다. 그 한 예가 그림 **1·9**에 있다. 그림 **1·9**에서는 연성을 고려하여 해석을 시행한 후 각 분야에서 정의된 최적화문제를 푼다. 그런데 각 분야의 반응함수 및 그의 민감도를 계산하기 위해 연성이 있는 해석을 다시 시행해야 하는 단점이 있다. 이를 극복하기 위해 각 분야의 최적화를 독립적으로 시행한 후에 연성된 해석을 다시 수행하는 등의 방법을 생각해 볼 수 있다. 이렇듯 다분야통합최적설계문제는 그 문제의 특성에 맞게 여러 가지 해법을 생각한 수 있다. 다뷰야통합최적설계는 최근에 대두된 이론으로서 앞으로 많은 연구가 필요하다.

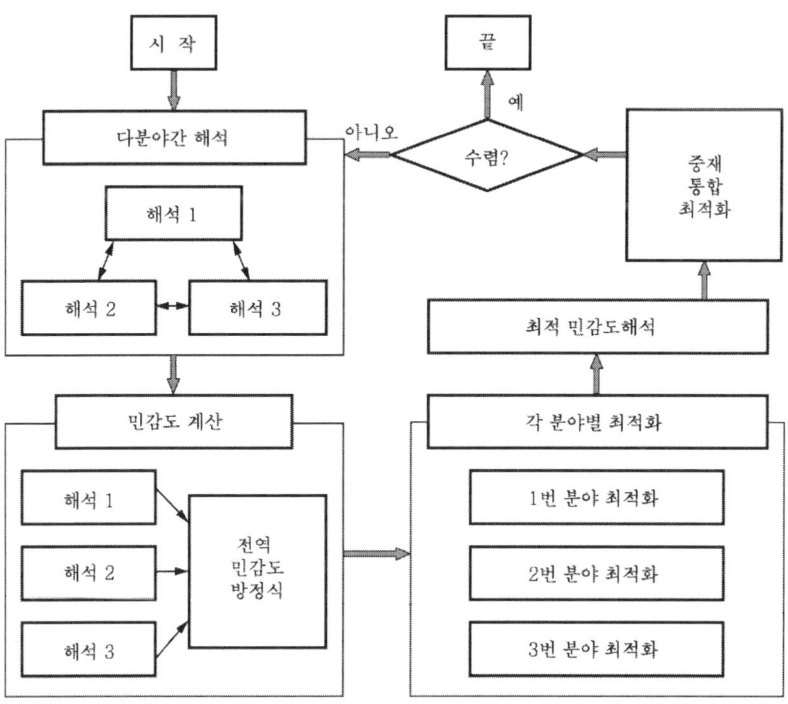

그림 1·9 다분야통합최적화의 문제 정의의 예

1.11 요 약

이상으로 이 책에서 취급하려는 각종 설계방법론을 간단히 설명하였다. 앞서 언급한 바와 같이 개념설계에서 유용하게 사용하는 이론으로 공리적설계방법을 제시하였다. 한편 설계프로세스를 효율적으로 증진시키는 방법으로 조립을 위한 설계법(design for assembly, DFA), 제조를 위한 설계법(design for manufacturing, DFM) 등의 여러 방법이 있다. 또한 앞서 언급했지만 설계 아이디어의 창출을 돕기 위한 TRIZ 등의 기법이 있다. 이들 역시 주요 개념설계방법들이기는 하지만 이 책에서는 생략한다.

상세설계에서 사용할 수 있는 설계법으로는 최적설계방법을 중심으로 실험계획법, 강건설계법을 간단히 설명하였다. 최근에는 이 외에도 많은 종류의 설계 알고리즘이 개발되었다. 예를 들어 유전자알고리즘(genetic algorithm, GA), 시뮬레이티드 어닐링(simulated annealing, SA), 퍼지(fuzzy)이론을 이용한 설계법, 신경회로망(neural network)을 이용한 설계법, 입자무리최적알고리즘(particle swarm optimization algorithm) 등이 그것들이다. 이들은 최적설계의 방법들이다. 이들에 대한 설명은 각 알고리즘을 전문적으로 다루는 문헌들을 참고하기 바란다.

참 고 문 헌

류연선 · 임오강 · 박경진 역, 『최적설계입문』, 반도출판사, 1994.

박성현, 『현대실험계획법』, 민영사, 1996.

차성운 · 박경진 역, 『공리적설계 [1]』, 동명사, 2002.

Altshuller G, *And Suddenly the Inventor Appeared*, Technical Innovation Center, Massachusetts, 1996.

Antonsson EK & Cagan J, *Formal Engineering Design Synthesis*, Cambridge University Press, New York, 2001.

Arora JS, *Guide to Structural Optimization*, The American Society of Civil Engineers, New York, 1997.

Arora JS, *Introduction to Optimum Design*, 2nd ed., Elsevier, CA, 2004.

Braun R, "*Collaborative Optimization: An Architecture for Large-Scale Distributed Design*," Ph. D. Thesis, Stanford University, California, 1996.

Choi WS & Park GJ, "Structural Optimization Using Equivalent Static Loads at All Time Intervals," *Computer Methods in Applied Mechanics and Engineering* 191, pp. 2077~2094, 2002.

Dieter GE, *Engineering Design: A Materials and Processing Approach*, 3rd ed., McGraw-Hill, New York, 1999.

Haftka RT & Gürdal, Z, *Elements of Structural Optimization*, 3rd Rev. and Expanded ed., Kluwer Academic Publishers, Dordrecht, 1992.

Haugen EB, *Probabilistic Mechanical Design*, John Wiley and Sons, New York, 1980.

Hibbeler RC, *Mechanics of Materials*, 4th ed., Prentice Hall, Eaglewood Cliffs, NJ, 2000.

Hurst K, *Engineering Design Principles*, Wiley, New York, 1999.

Hyman BI, *Fundamentals of Engineering Design*, Prentice Hall, Eaglewood Cliffs, NJ, 1998.

Kang BS & Park GJ & Arora JS, "A Review of Optimization of Structures Subjected to Trasient Loads," *Structural and Multidisciplinary Optimization* 31, pp. 81~95, 2006.

Montgomery DC, *Design and Analysis of Experiments*, Wiley, New York, 1997.

Norton RL, *Machine Design: An Integrated Approach*, 2nd ed., Prentice Hell, Eaglewood Cliffs, NJ, 2000.

Park GJ & Kang BS, "Validation of Structural Optimization Algorithm Transforming Dynamic Loads into Equivalent Static Loads," *Journal of Optimization Theory and Applications* 119(1), pp. 191~200, 2002.

Park GJ & Lee TH & Lee KH & Hwang KH, "Robust Design: An Overview," *AIAA Journal* 44(1), pp. 181~191, 2006.

Phadke MS, *Quality Engineering Using Robust Design*, Prentice Hall, New Jersey, 1989.

Rao SS, *Engineering Optimization: Theory and Practice*, 3rd ed., Wiley, New York, 1996.

Sobieszcznski-Sobieski J & Haftka RT, "Multidisciplinary Aerospace Design Optimization: Survey of Recent Developments," *Structural and Multidisciplinary Optimization* 14(1), pp. 1~23, 1997.

Suh NP, *The Principles of Design*, Oxford University Press, New York, 1990.

Suh NP, *Axiomatic Design: Advances and Applications*, Oxford University Press, New York, 2001.

Taguchi G, *System of Experimental Design* Vol. 1 · 2, Kraus, New York, 1987.

Taylor WA, *Optimization and Variation Reduction in Quality*, McGraw-Hill, New York, 1991.

Ullman DG, *The Mechanical Design Process*, 2nd ed., McGraw-Hill, New York, 1997.

제 2 장

공리적설계

2.1 공리적설계

공리적설계는 매사추세츠 공과대학의 서남표 교수에 의해 창출되고 보급되어 온 설계 이론이다(Suh, 1990 · 2000). 그러나 좀더 정확히 말하면 공리적설계는 설계이론이라기보다는 일반적인 공학적 과정에 적용할 수 있는 원리를 포함하는 설계프레임워크라 할 수 있다. 공리적설계는 독립공리와 정보공리로 이루어져 있다. 공리적설계에서 좋은 설계는 두 공리를 만족하는 설계이다. 두 공리를 만족하지 않는 설계는 좋은 설계가 아니다. 공리라는 용어는 기하학에서 사용되는 것과 동일한 의미로, 증명되지는 않지만 반례가 발견될 경우 거짓이 된다는 것을 의미한다. 그런데 아직까지는 공리적설계의 공리들에 대한 반례가 제시되지 않았다. 오히려 공리적설계로 설명이 가능한 좋은 설계의 예들이 증가하고 있다.

설계는 '무엇을 이룰 것인가?'와 '그것을 어떻게 얻을 것인가?'의 상호작용이라 정의할 수 있다. 설계자는 양쪽의 기능 및 자원을 적절히 조합하고 배치해서 원하는 것을 얻으려 한다. 설계를 수행할 때에 문제를 정의하고 설계를 완성하는 등의 순서에 따라 그림 **2 · 1**과 같은 영역들이 있다. 소비자 영역에서는 소비자들이 원하는 요구를 정의하며 대개 이들 요구는 서술적으로 표현할 수 있다. 그러나 최근에는 소비자들의 요구가 없는데도 소비자가 필요로 할 것 같은 사항을 생산자가 예측하여 직접 정의하는 수도 있다. 이렇게 정의한 것을 소비자요구(customer attributes, CAs)라 한다. 소비자 영역에서 정의한 사항은 기능적 영역에서 기능요구(functional requirements, FRs)로 변환된다. 기능요구는 소비자 영역의 소비자요구를 공학적 용어로 재정의한 것이다. 이는 앞에서 언급한 '무엇을 이룰

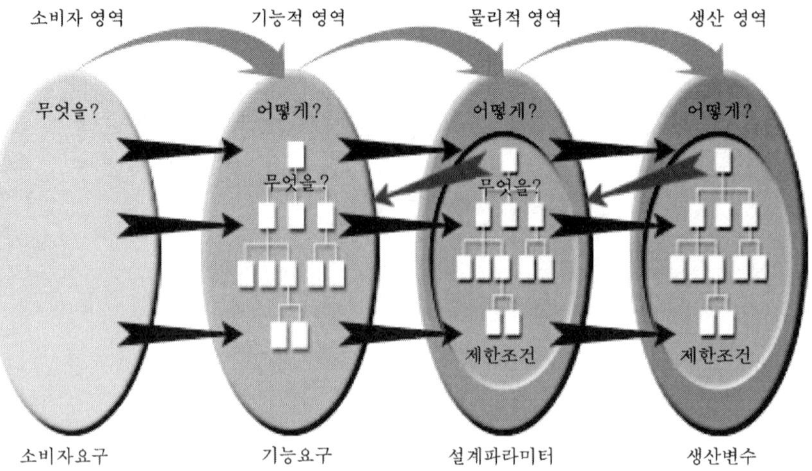

그림 2 · 1 도메인, 사상, 영역의 관계

것인가?'에 대한 공학적 정의라 볼 수 있다. 기능요구는 물리적 영역에서 설계파라미터(design parameters, DPs)를 정의하거나 선정함으로써 만족시킨다. 대개 이 과정을 설계과정이라 부른다. 설계파라미터를 결정하면 위와 동일한 방법으로 설계파라미터로부터 제조를 위한 생산변수(production variables, PVs)를 정의하게 된다. 이와 같이 공리적설계에서의 설계는 두 영역 사이의 관계를 파악하여 앞의 영역에서 정의한 사항에 따라 다음 영역에서의 사항을 정의한다. 이를 영역간 사상이라 한다. 좋은 설계란 이러한 사상을 효율적으로 정의한 것을 뜻하기도 한다.

설계공리는 모든 공학적 행위의 근간에 흐르는 공통적인 원리를 집약적으로 표현한 것으로 다음과 같이 정의한다(NSF, 1998 ; Suh, 1990).

제 1 공리 : 독립공리(The Independence Axiom)
기능요구의 독립성을 유지하라 (Maintain the independence of the functional requirements(FRs)).

제 2 공리 : 정보공리(The Information Axiom)
설계 대상의 정보량을 최소화하라 (Minimize the information content of the design).

공리는 대개 기하학에서 정의해 왔는데, 기하학의 경우와 마찬가지로 여기에서도 공리로부터 증명이 가능한 많은 정리 및 추론을 도출할 수 있다. 현재까지의 각종 정리 및 추론이 부록 **2.A**에 수록되어 있다. 여기서는 증명은 생략한다.

2.2 독립공리

2.2.1 독립공리

독립공리는 그림 **2 · 1**에 있는 영역간의 사상 과정에서 오른쪽 영역에 정의되는 사항들이 왼쪽 영역의 사항들을 독립적으로 만족시키는 것이 좋은 설계라는 의미이다. 앞 절에서 정의한 FR-DP 혹은 DP-PV간의 관계를 독립적으로 유지하라는 것이다. 복수의 FR이 정의되었을 때 각 DP가 각 FR을 독립적으로 만족시킬 수 있어야 한다. 이러한 FR-DP간의 관계는 설계행렬로 표현할 수 있다. 복수의 FR과 DP는 벡터 $\mathbf{FR}$과 $\mathbf{DP}$로 표시할 수 있다. 이들의 관계는 설계행렬 $\mathbf{A}$를 사용하여 다음과 같이 표현할 수 있다.

$$\mathrm{FR} = \mathrm{A\ DP} \tag{2・1}$$

$\mathbf{FR}$과 $\mathbf{DP}$ 벡터의 요소 수가 각각 3이라 하면 행렬 $\mathbf{A}$는 다음과 같다.

$$\begin{bmatrix} FR_1 \\ FR_2 \\ FR_3 \end{bmatrix} = \begin{bmatrix} A_{11} & A_{12} & A_{13} \\ A_{21} & A_{22} & A_{23} \\ A_{31} & A_{32} & A_{33} \end{bmatrix} \begin{bmatrix} DP_1 \\ DP_2 \\ DP_3 \end{bmatrix} \tag{2・2}$$

설계행렬 $\mathbf{A}$의 특성에 따른 관계식이 표 **2 · 1**에 보인다. 표 **2 · 1**에서 설계행렬이 대각행렬이 되는 경우를 비연성설계(uncoupled design)라 한다. 이 경우 각 설계파라미터가 대응하는 기능요구를 독립적으로 만족할 수 있으므로 완벽하게 독립공리를 만족하게 된다. 다음으로 설계행렬이 삼각행렬인 경우가 있다. 이것을 비연성화설계(decoupled design)라 부른다. 비연성화설계는 설계파라미터를 결정하는 순서에 따라서 기능요구를 독립적으로

표 2 · 1 설계행렬의 특성에 따른 FR-DP 관계

	설계방정식	설계 결과
비연성설계	$\begin{bmatrix} FR_1 \\ FR_2 \\ FR_3 \end{bmatrix} = \begin{bmatrix} A_{11} & 0 & 0 \\ 0 & A_{22} & 0 \\ 0 & 0 & A_{33} \end{bmatrix} \begin{bmatrix} DP_1 \\ DP_2 \\ DP_3 \end{bmatrix}$	$FR_1 = A_{11} \times DP_1$ $FR_2 = A_{22} \times DP_2$ $FR_3 = A_{33} \times DP_3$
비연성화설계	$\begin{bmatrix} FR_1 \\ FR_2 \\ FR_3 \end{bmatrix} = \begin{bmatrix} A_{11} & 0 & 0 \\ A_{21} & A_{22} & 0 \\ A_{31} & A_{32} & A_{33} \end{bmatrix} \begin{bmatrix} DP_1 \\ DP_2 \\ DP_3 \end{bmatrix}$	$FR_1 = A_{11} \times DP_1$ $FR_2 = A_{21} \times DP_1 + A_{22} \times DP_2$ $FR_3 = A_{31} \times DP_1 + A_{32} \times DP_2 + A_{33} \times DP_3$
연성설계	$\begin{bmatrix} FR_1 \\ FR_2 \\ FR_3 \end{bmatrix} = \begin{bmatrix} A_{11} & A_{12} & A_{13} \\ A_{21} & A_{22} & A_{23} \\ A_{31} & A_{32} & A_{33} \end{bmatrix} \begin{bmatrix} DP_1 \\ DP_2 \\ DP_3 \end{bmatrix}$	$FR_1 = A_{11} \times DP_1 + A_{12} \times DP_2 + A_{13} \times DP_3$ $FR_2 = A_{21} \times DP_1 + A_{22} \times DP_2 + A_{23} \times DP_3$ $FR_3 = A_{31} \times DP_1 + A_{32} \times DP_2 + A_{33} \times DP_3$

만족시킬 수 있다. 표 **2·1**의 두 번째 경우를 보면 우선 DP_1을 결정하여 FR_1을 만족시킨다. 이어서 DP_2, DP_3을 결정하여 FR_2, FR_3을 순서대로 만족시키게 된다. 그러나 표 **2·1**의 세 번째의 경우에서는 어떤 순서로도 FR과 DP가 연성의 관계를 가진다. 이를 연성설계(coupled design)라 한다. 비연성설계나 비연성화설계는 독립공리를 만족하는 것으로 간주하며 연성설계는 독립공리를 위배하는 것으로 간주한다. 위에서 기능요구인 FR과 설계파라미터인 DP를 표기할 때 각 요소들을 구분하기 위해 아래 첨자를 사용했다. 그러나 표현방법에 있어서 아래첨자를 사용하지 않기도 한다. 따라서 FR_i와 FRi는 동일한 것이다. 한편 설계행렬 A는 일반적인 수학적 연산 가운데 곱셈만 가능하고 좌표변환 등의 연산은 불가능하다.

어떤 설계를 공리적설계방법으로 분석했을 때, 비연성설계나 비연성화설계는 좋은 설계이며 연성설계는 그렇지 않다라는 것을 의미한다. 따라서 연성설계는 설계파라미터의 재조정을 통해 비연성설계나 비연성화설계로 변환하여야 한다.

여기서 주목할 사항은 제한조건(constraints, Cs)이다. 제한조건은 설계 시 반드시 만족시켜야 하는 항목이다. 예컨대 비용이나 설계조건 등은 어떤 환경에서도 만족되어야 하는 명제이다. 제한조건은 설계의 독립성에 관계없이 정의할 수 있으며 다수의 설계파라미터가 연성되어 있을 수도 있다. 그림 **2·1**에서 보는 바와 같이 DP나 PV영역에서 제한조건을 정의할 수 있다.

다음의 예제에서 독립공리의 응용을 생각해 본다. 예제에서 보는 바와 같이 일반적으로 FR의 표현에는 명령문을, DP에는 명사를 사용한다.

예제 2·1 냉장고 문 설계 문제 (NSF, 1998 ; Suh, 2000)

자주 접할 수 있는 냉장고의 문은 그림 **2·2**에 보는 바와 같이 두 가지이다. 이 중 어떤 것이 우수한 설계일까? 이에 대한 해답을 찾기 위해 냉장고 문의 기능요구를 다음과 같이 정의한다.

FR_1 : 냉장고 안의 음식에 접근할 수 있도록 하여라.

FR_2 : 에너지 손실을 최소화하여라.

■ ‖ 풀 이 ‖ ■

그림 **2·2** (a)에 있는 수직문의 설계파라미터를 다음과 같이 설정할 수 있다.

DP_1 : 수직문

DP_2 : 문에 설치한 단열재

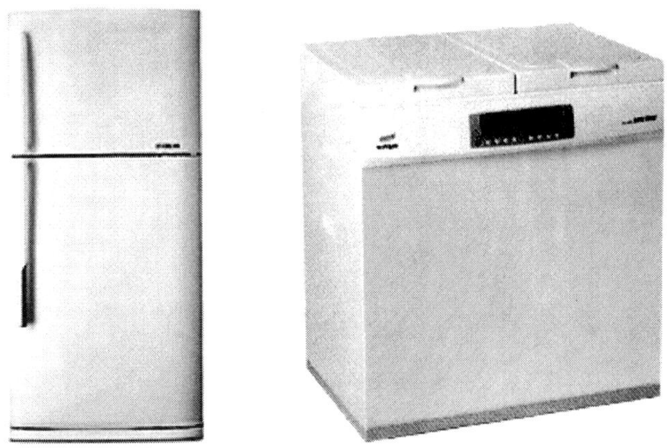

(a) 수직문　　　　(b) 수평문

그림 2 · 2 냉장고 문의 모습

설계행렬을 이용한 그들의 관계는 다음과 같다.

$$\begin{bmatrix} FR_1 \\ FR_2 \end{bmatrix} = \begin{bmatrix} X & 0 \\ X & X \end{bmatrix} \begin{bmatrix} DP_1 \\ DP_2 \end{bmatrix} \tag{2 · 3}$$

여기서 X는 영향이 있음을, 0은 영향이 없음을 나타낸다.

이상에서 수직문의 설계는 비연성화설계임을 알 수 있다. 따라서 독립공리를 만족하는 설계이다. 그러나 비대각항의 X로 인해 문을 열 때 열손실이 발생하고 있다. 이를 극복하는 것으로 최근 김치 냉장고 등에 있는 수평문을 보자. 그림 **2 · 2** (b)에 보여지는 수평문의 공리적설계에 의한 분석은 다음과 같다.

DP_1 : 수평문

DP_2 : 문에 설치한 단열재

$$\begin{bmatrix} FR_1 \\ FR_2 \end{bmatrix} = \begin{bmatrix} X & 0 \\ 0 & X \end{bmatrix} \begin{bmatrix} DP_1 \\ DP_2 \end{bmatrix} \tag{2 · 4}$$

수평문의 경우 문을 열 때 찬공기가 그대로 남아 있어서 열손실을 최소화할 수 있다. 따라서 수평문은 비연성설계이고 수직문보다 우수한 설계이다. 그러면 수평문을 가진 냉장고가 항상 우수할까? 그 대답은 이 예제의 기능요구에 대해서는 '그렇다'이다. 그러나 저장할 음식물의 양이나 음식에 접근하는 편리성 등을 제한조건으로 제시하면 문제는 달라진다. 수평문의 냉장고가 충분한 저장량을 보유하지 못하거나 음식물에의 접근이 어려워 제한조건을 위배한다면, 독립성 여부와 관계없이 채택할 수 없는 설계가 되는 것이다. 제한조건이 있을 때는 우선적으로 제한조건을 만족하는 설계 중 독립공리의 만족 여부를 판단해야 한다.

예제 2 · 2　수도꼭지의 설계　(Suh, 2000)

더운물과 찬물이 사용자의 의도에 따라 용이하게 배합되는 수도꼭지의 설계를 생각해 본다. 이미 상품화된 것이 많으므로 기존의 제품을 가지고 평가한다. 수도꼭지의 기능요구는 다음과 같이 정의한다.

FR_1 : 수도꼭지에서 나오는 물의 양(Q)을 조절하여라.

FR_2 : 수도꼭지에서 나오는 물의 온도(T)를 조절하여라.

■|| 풀 이 ||■

그림 **2 · 3** (a)에 보이는 제품을 분석하면 다음과 같이 설계파라미터와 설계행렬이 만들어진다.

DP_1 : 그림 **2 · 3** (a)의 각도 ϕ_1

DP_2 : 그림 **2 · 3** (a)의 각도 ϕ_2

$$\begin{bmatrix} FR_1(Q) \\ FR_2(T) \end{bmatrix} = \begin{bmatrix} X & X \\ X & X \end{bmatrix} \begin{bmatrix} DP_1(\phi_1) \\ DP_2(\phi_2) \end{bmatrix} \tag{2 · 5}$$

식 (2 · 5)의 설계행렬에서 보듯이, 그림 **2 · 3** (a) 제품의 설계는 연성되어 있다. 따라서 좋지 않은 설계라는 결론이 뒤따른다.

다른 예로서 그림 **2 · 3** (b)를 들 수 있다. 이에 대한 분석은 다음과 같다.

DP_1 : 그림 **2 · 3** (b)의 각도 ϕ_1

DP_2 : 그림 **2 · 3** (b)의 각도 ϕ_2

$$\begin{bmatrix} FR_1(Q) \\ FR_2(T) \end{bmatrix} = \begin{bmatrix} X & 0 \\ 0 & X \end{bmatrix} \begin{bmatrix} DP_1(\phi_1) \\ DP_2(\phi_2) \end{bmatrix} \tag{2 · 6}$$

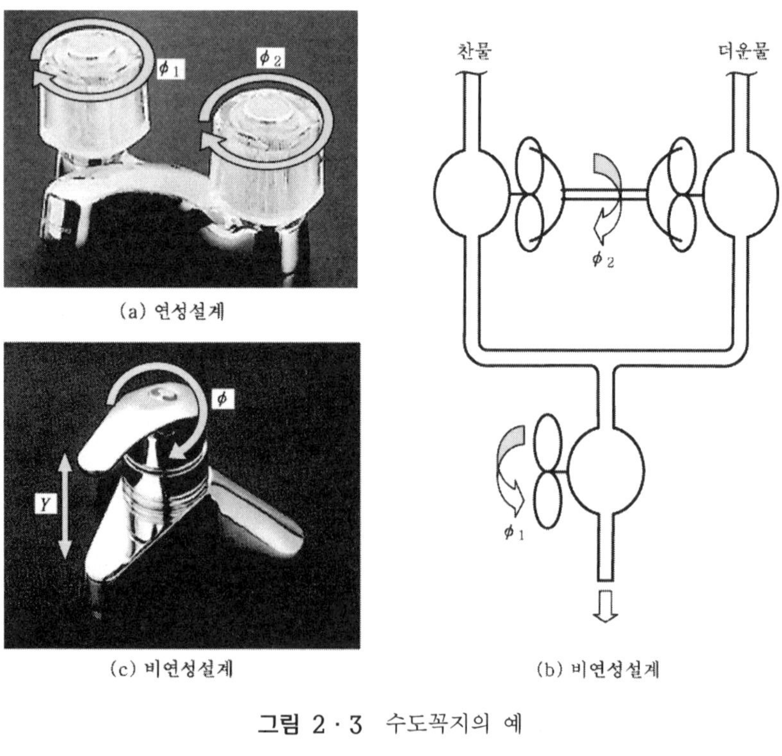

그림 2 · 3 수도꼭지의 예

설계행렬이 대각행렬이므로 비연성설계이다. 따라서 독립공리를 만족하며 좋은 설계라 할 수 있다.

또 다른 예로서 그림 **2 · 3** (c)의 경우를 보자. 분석 결과는 다음과 같다.

DP_1 : 그림 **2 · 3** (c)의 변위 Y

DP_2 : 그림 **2 · 3** (c)의 각도 ϕ

$$\begin{bmatrix} FR_1(Q) \\ FR_2(T) \end{bmatrix} = \begin{bmatrix} X & 0 \\ 0 & X \end{bmatrix} \begin{bmatrix} DP_1(Y) \\ DP_2(\phi) \end{bmatrix} \qquad (2 \cdot 7)$$

설계행렬이 대각행렬이므로 독립공리의 관점에서 볼 때 그림 **2 · 3** (b)(c)는 우수한 설계라 볼 수 있다. 그러면 그림 **2 · 3** (b)와 **2 · 3** (c) 중 어느 것이 더 좋은 설계일까? 그림 **2 · 3** (c)의 설계는 조작이 간편한 반면 그림 **2 · 3** (b)는 조작이 대체로 복잡하다. 따라서 간편한 그림 **2 · 3** (c)가 가장 우수한데 이는 나중에 언급할 정보공리로 설명할 수 있다. 실제로 그림 **2 · 3** (c)의 설계가 최근 급속도로 보급되고 있는 것이 이를 대변해 준다고 볼 수 있다. 그러나 이는 공학적인 기능에 관한 것이며 미적 사항을 중요하게 여기는 설계에서는 다른 판단을 할 수도 있다.

복잡한 시스템을 설계할 때 단지 한번의 FR-DP의 관계 설정만으로는 부족할 수가 있다. 이런 경우에는 일단 설정된 FRs와 DPs를 다시 분해하게 된다. 그림 **2 · 4**에 보는 바와 같이 기능적 영역과 물리적 영역을 지그재그로 오가며 FR-DP관계를 정의한다. 이 지그재그 과정을 그림 **2 · 4**에서 번호순으로 보이고 있다. 여기서 주목할 점은 DP는 같은 단계의 FR에 의해 정의되고, FR은 위 단계의 DP의 특성에 따라 아래 단계로 분해된다는 것이다. 선반의 설계를 공리적설계의 관점에서 여러 단계로 분해한 예가 그림 **2 · 5**에 있다(Suh 1990).

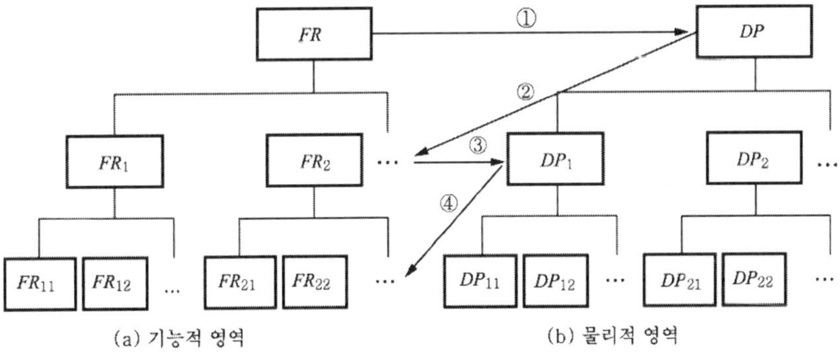

그림 2 · 4 영역간 지그재그 과정

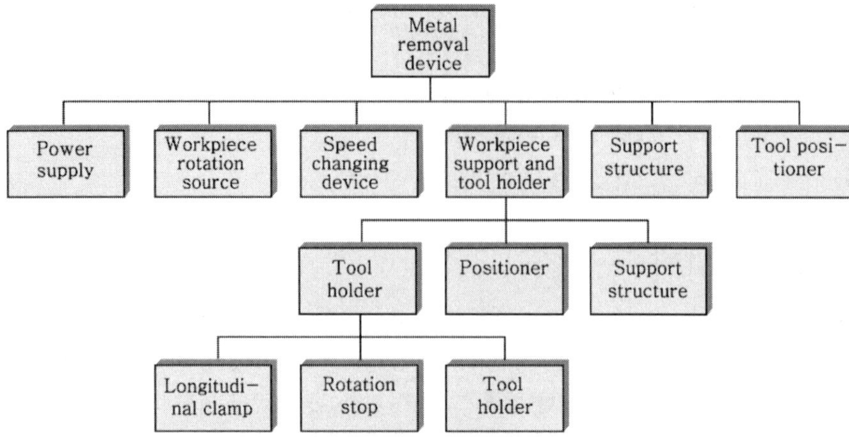

(a) 기능적 영역

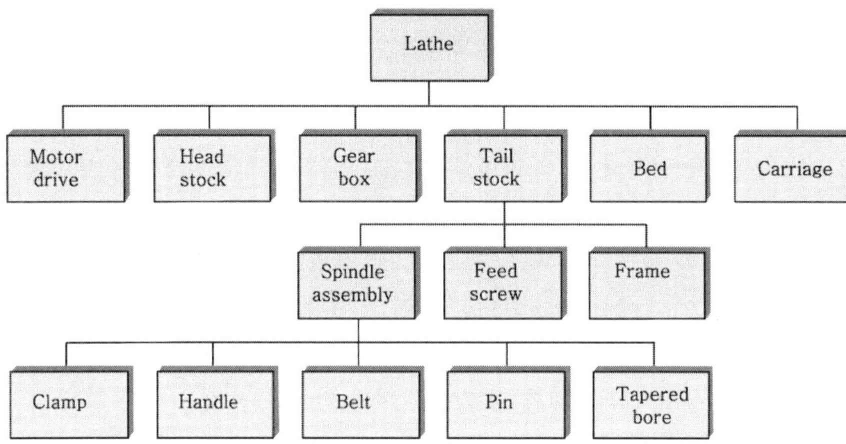

(b) 물리적 영역

그림 2 · 5 공리적설계방법에 의한 선반 설계의 분석

2.2.2 독립성

기능요구와 설계파라미터 간의 관계는 식 (2 · 1)로 나타낸다. 기능요구와 설계파라미터의 개수가 2일 때, 설계행렬의 특성에 따라 사상 과정을 FR과 DP좌표계의 중첩을 통해 도식화한 것이 그림 **2 · 6**에 있다. 각각의 설계는 다음과 같은 설계행렬로 나타낼 수 있다(Rinderle, 1982).

$$\begin{bmatrix} FR_1 \\ FR_2 \end{bmatrix} = \begin{bmatrix} A_{11} & 0 \\ 0 & A_{22} \end{bmatrix} \begin{bmatrix} DP_1 \\ DP_2 \end{bmatrix} \quad \text{비연성설계(그림 2 · 6 (a))} \qquad (2 \cdot 8)$$

$$\begin{bmatrix} FR_1 \\ FR_2 \end{bmatrix} = \begin{bmatrix} A_{11} & A_{12} \\ 0 & A_{22} \end{bmatrix} \begin{bmatrix} DP_1 \\ DP_2 \end{bmatrix} \quad \text{비연성화설계(그림 2 · 6 (b))} \qquad (2 \cdot 9)$$

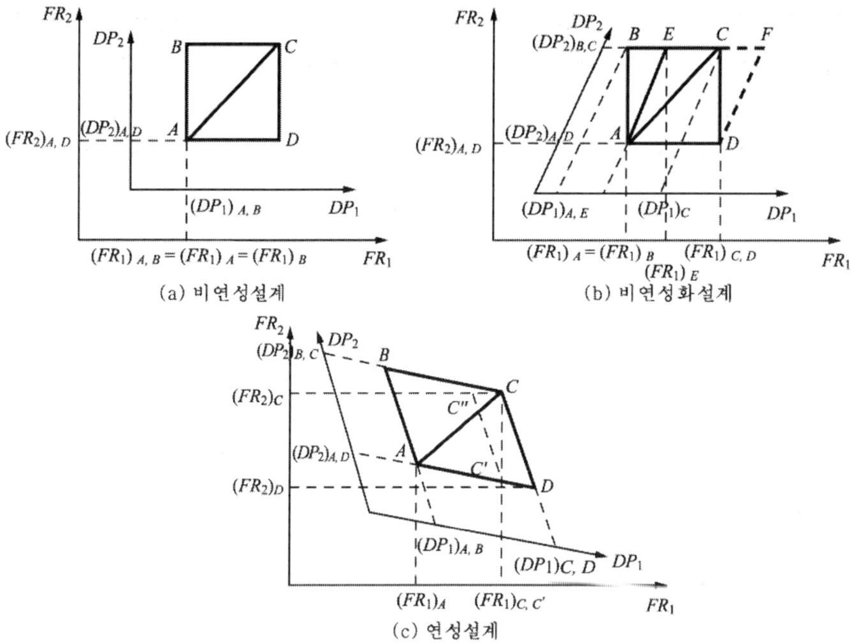

그림 2 · 6 각 설계에 대한 FR영역으로부터 DP영역에로의 사상 과정의 도식화

$$\begin{bmatrix} FR_1 \\ FR_2 \end{bmatrix} = \begin{bmatrix} A_{11} & A_{12} \\ A_{21} & A_{22} \end{bmatrix} \begin{bmatrix} DP_1 \\ DP_2 \end{bmatrix} \quad \text{연성설계(그림 2 · 6 (c))} \qquad (2 \cdot 10)$$

그림 **2 · 6** (a)에 있는 비연성설계를 보자. 상태 A는 $(DP_1)_A$, $(DP_2)_A$에 대해 $(FR_1)_A$, $(FR_2)_A$의 값을 가지며 상태 B, C, D도 동일한 성격을 지닌다. 상태 A에서 C로 설계를 변경하려면 A-D-C를 거쳐도 되고 A-B-C를 거쳐도 된다. 그러므로 비연성설계는 경로에 무관한 설계이다. 그러나 그림 **2 · 6** (b)의 비연성화설계는 다소 다른 특성을 보인다. 가령 상태 A에서 상태 C로 변경하려면 일단 $(DP_2)_A$에서 $(DP_2)_E$로 DP_2만 변화시킨다. 이때 DP_1은 고정되어 있으며 FR_1과 FR_2는 모두 변한다. 이후에 DP_2를 고정하고 DP_1을 $(DP_1)_E$에서 $(DP_1)_C$로 변화시키면 된다. 이 과정에서는 FR_2는 고정되어 있고 FR_1만이 변하게 된다. 이와 같이 비연성화설계는 경로에 의존하는 설계가 된다. 즉, DP_2를 먼저 결정하고 DP_1을 나중에 결정하는 순서를 따라야 한다. 그러면 그림 **2 · 6** (c)의 연성설계를 보자. 상태 A에서 상태 C로 변하려 할 때, 설계파라미터의 변화 순서를 어떻게 하더라도 효과는 마찬가지이다. DP_1을 먼저 변화시켜 보자. FR_1을 만족하기 위해서 DP_1을 A에서 C'으로 변화시켜 보면 FR_2도 변함을 알 수 있다. FR_2를 만족시키기 위해서 DP_2를 변화시키면 C'에서 C''으로 움직인다. 그러면 FR_1도 변하게 되어 다시 DP_1을 변

화시켜야 된다. 따라서 설계를 만족하기 위해서는 수렴할 때까지 이러한 반복 과정을 거치게 된다. 이것은 매우 복잡해질 수도 있으며 설계행렬이 비선형일 때에는 해의 수렴이 불가능할 수도 있다.

위의 관계를 간략히 보인 것이 그림 **2·7**에 있다. 결국 설계행렬의 특징은 그림 **2·7**의 α_1, α_2, θ로 나타낼 수 있다. 이상적인 설계인 비연성설계는 $\alpha_1 = \alpha_2 = 0$이고 $\theta = 90°$일 경우이다. 연성의 정도를 나타내는 지수로 식 (2·11)로 정의되는 리앵귤러리티(reangularity)라는 지수 R을 사용한다.

$$R = \sin\theta = (1 - \cos^2\theta)^{\frac{1}{2}} \tag{2·11}$$

FR과 DP의 개수가 n이고 설계행렬의 요소를 A_{ij}이라 할 때 일반적인 R은 다음과 같이 표시된다.

$$R = \prod_{\substack{i=1,\ n-1 \\ j=1+i,\ n}} \left[1 - \frac{\left(\sum_{k=1}^{n} A_{ki} A_{kj} \right)^2}{\left(\sum_{k=1}^{n} A_{ki}^2 \right) \left(\sum_{k=1}^{n} A_{kj}^2 \right)} \right]^{\frac{1}{2}} \tag{2·12}$$

여기서 $\theta = 90°$일 때, 즉 그림 **2·7**에서 DP_1축과 DP_2축이 직교할 때 $R = 1$이다.

리앵귤러리티 R만 가지고는 연성의 정도를 모두 나타낼 수는 없다. 왜냐하면 $R = 1$이 된다고 해도 $\alpha_1 \to 0$과 $\alpha_2 \to 0$을 보장하지는 않기 때문이다. α_1과 α_2가 영이 된다는 것은 설

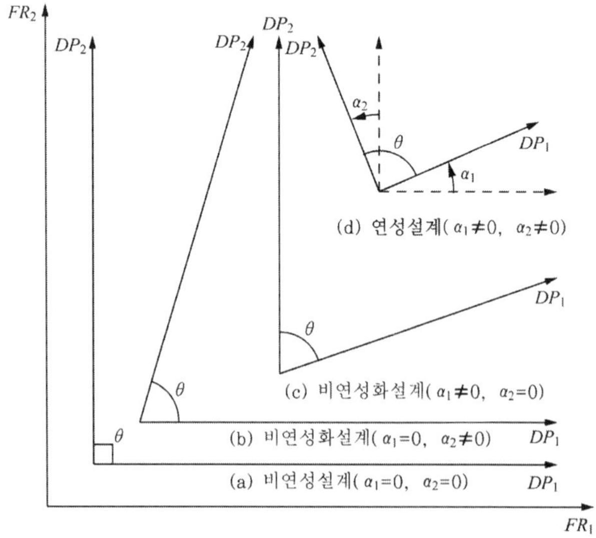

그림 2·7 연성 특성에 따른 설계의 도식화

계행렬이 대각행렬이 되는 것을 의미하며, 행렬의 대각 항이 상대적으로 클 때 연성 정도가 낮게 된다. 여기서 주목할 점은 α_1이나 α_2 중 하나가 0이면 R이 0에 가깝더라도 비연성화설계가 된다. 이에 대한 지수로 다음의 셈앵귤러리티(semangularity, 라틴말로 same angle quality를 의미한다) S를 정의한다.

$$S = \prod_{j=1}^{n} \frac{|A_{jj}|}{\left(\sum_{k=1}^{n} A_{kj}^2\right)^{\frac{1}{2}}} \tag{2·13}$$

설계행렬이 대각행렬일 때, 즉 $\alpha_1 = \alpha_2 = 0$일 때 $S = 1$의 값을 가진다. 각 설계에 대한 리앵귤러리티와 셈앵귤러리티의 특성이 표 **2·2**에 정리되어 있다. 비연성화설계에서는 R과 S값이 1보다는 작으나 동일한 값을 가지며 그 외의 경우는 연성설계이다.

표 2 2 각 설계에 대한 리앵귤러리티와 셈앵귤러리티의 특성

	비연성설계	비연성화설계	연성설계
리앵귤러리티	1	$R = S < 1$	$R \neq S < 1$
열벡터 간의 각도(θ)	90°	θ	θ
셈앵귤러리티	1	$S = R < 1$	$S \neq R < 1$

예제 2 · 3 비연성화설계에서의 리앵귤러리티와 셈앵귤러리티

식 (2·9)의 비연성화설계에서 R과 S값이 동일함을 증명하여라.

풀 이 1

기능요구가 두 개일 경우의 R과 S의 값을 각각 식 (2·12)와 (2·13)을 이용하여 다음과 같이 얻을 수 있다.

$$R = \left[1 - \frac{(A_{11}A_{12} + A_{21}A_{22})^2}{(A_{11}^2 + A_{21}^2)(A_{12}^2 + A_{22}^2)}\right]^{\frac{1}{2}} \tag{2·14a}$$

$$S = \left[\frac{|A_{11}|}{\sqrt{A_{11}^2 + A_{21}^2}}\right]\left[\frac{|A_{22}|}{\sqrt{A_{12}^2 + A_{22}^2}}\right] \tag{2·14b}$$

식 (2·9)에서 $A_{21} = 0$이므로 이를 식 (2·14)에 대입하면 다음과 같다.

$$R = \left[1 - \frac{A_{11}^2 A_{12}^2}{A_{11}^2(A_{12}^2 + A_{22}^2)}\right]^{\frac{1}{2}} = \left[\frac{A_{11}^2 A_{22}^2}{A_{11}^2(A_{12}^2 + A_{22}^2)}\right]^{\frac{1}{2}} = \frac{|A_{22}|}{\sqrt{A_{12}^2 + A_{22}^2}} \tag{2·15a}$$

$$S=\left[\frac{|A_{11}|}{\sqrt{A_{11}^2}}\right]\left[\frac{|A_{22}|}{\sqrt{A_{12}^2+A_{22}^2}}\right]=\frac{|A_{22}|}{\sqrt{A_{12}^2+A_{22}^2}} \tag{2·15b}$$

그러므로 R과 S는 동일하다.

▪||풀 이2||▪

위의 문제를 기하학적으로 풀어 본다. $\mathbf{c}_1=[A_{11}\ \ 0]^T$, $\mathbf{c}_2=[A_{12}\ \ A_{22}]^T$라 하면 두 개의 기능요구를 벡터 FR로 정의할 때 $\mathrm{FR}=DP_1\mathbf{c}_1+DP_2\mathbf{c}_2$가 되는데 이에 대한 기하학적인 해석이 그림 2·8에 있다. 이로부터 R과 S의 값을 구하면 다음과 같다.

$$\cos\theta=\frac{\mathbf{c}_1^T\mathbf{c}_2}{|\mathbf{c}_1||\mathbf{c}_2|}=\frac{A_{11}A_{12}}{\sqrt{A_{11}^2}\sqrt{A_{12}^2+A_{22}^2}} \tag{2·16a}$$

$$R=\sin\theta=(1-\cos^2\theta)^{\frac{1}{2}}=\left[1-\frac{A_{11}^2A_{12}^2}{A_{11}^2(A_{12}^2+A_{22}^2)}\right]^{\frac{1}{2}}$$

$$=\left[\frac{A_{11}^2A_{22}^2}{A_{11}^2(A_{12}^2+A_{22}^2)}\right]^{\frac{1}{2}}=\frac{|A_{22}|}{\sqrt{A_{12}^2+A_{22}^2}} \tag{2·16b}$$

$$S=\cos\alpha_1\cos\alpha_2=\left[\frac{|A_{11}|}{\sqrt{A_{11}^2+0^2}}\right]\left[\frac{|A_{22}|}{\sqrt{A_{12}^2+A_{22}^2}}\right]=\frac{|A_{22}|}{\sqrt{A_{12}^2+A_{22}^2}} \tag{2·16c}$$

따라서 R과 S가 동일함을 알 수 있다.

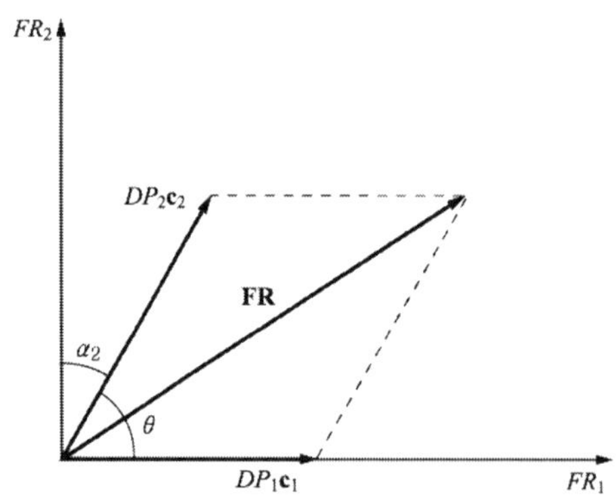

그림 2·8 예제 2·3의 도식화

설계행렬이 설계파라미터에 대해 비선형일 경우를 고려한다. 이때에는 식 (2·2)의 A_{ij}가 상수가 아니고 DP의 함수이다. 현재의 설계점을 DP_0이라 할 때 설계방정식은 $\mathrm{FR}_0=\mathrm{ADP}_0$이라 할 수 있다. 설계행렬의 각 요소를 구하기 위해 식 (2·2)를 테일러 급수로 전개하여 근사화 함수를 구하면 식 (2·2)의 FR_i는 다음과 같이 된다.

$$FR_i = (FR_i)_0 + \delta FR_i = (FR_i)_0 + \sum_{j=1}^{3} \frac{\partial FR_i}{\partial DP_j} \delta DP_j \qquad (2 \cdot 17)$$

설계행렬은 식 (2 · 17)을 이용하여 다음과 같이 현재의 설계점 DP_0에서 정의한다.

$$\begin{bmatrix} \delta FR_1 \\ \delta FR_2 \\ \delta FR_3 \end{bmatrix} = \begin{bmatrix} \frac{\partial FR_1}{\partial DP_1} & \frac{\partial FR_1}{\partial DP_2} & \frac{\partial FR_1}{\partial DP_3} \\ \frac{\partial FR_2}{\partial DP_1} & \frac{\partial FR_2}{\partial DP_2} & \frac{\partial FR_2}{\partial DP_3} \\ \frac{\partial FR_3}{\partial DP_1} & \frac{\partial FR_3}{\partial DP_2} & \frac{\partial FR_3}{\partial DP_3} \end{bmatrix} \begin{bmatrix} \delta DP_1 \\ \delta DP_2 \\ \delta DP_3 \end{bmatrix} \qquad (2 \cdot 18)$$

식 (2 · 18)에서 보는 바와 같이 현재의 설계점에서 기능요구와 설계파라미터가 증분으로 표시되며, 설계행렬의 각 항은 편미분 값이다. 비연성설계를 찾는 노력은 식 (2 · 18)의 설계행렬이 주어진 증분의 범위에서 대각행렬이 되는 설계창(design window)을 찾는 일이다. 그리고 한 점에서 비연성관계를 가져도 설계점이 바뀔 경우 이것이 지속된다는 보장은 없다.

여기서 주목할 또 하나의 사항은, 설계행렬의 비대각항이 대각항보다 그 크기가 현저히 작을 때 이를 비연성설계로 취급할 수 있느냐는 것이다. 다음 식을 보자.

$$\begin{bmatrix} FR_1 \\ FR_2 \\ FR_3 \end{bmatrix} = \begin{bmatrix} X & x & x \\ x & X & x \\ x & x & X \end{bmatrix} \begin{bmatrix} DP_1 \\ DP_2 \\ DP_3 \end{bmatrix} \qquad (2 \cdot 19)$$

여기서 $X \gg x$이다. 이때는 설계에서 주어진 공차범위와 X, x의 크기로 판단할 수 있다. 이에 대한 근거가 부록의 정리 **2.A · 8**에 있다. 정리 **2.A · 8**을 소개하면 n이 기능요구의 수라 할 때 다음과 같다.

【정리 2.A · 8】 독립성과 공차

설계자로부터 지정된 공차가 다음 식보다 크면 설계는 비연성설계이다.

$$\sum_{\substack{j \neq i \\ j=1}}^{n} \left(\frac{\partial FR_i}{\partial DP_j} \right) \Delta DP_j \qquad (2 \cdot 20)$$

여기서 ΔDP_j는 j번째 설계파라미터의 현재점에서 설계의 변화량, 즉 설계창의 크기이다. 이러한 경우에서 설계행렬의 비대각 원소들을 설계 고려사항에서 제외할 수 있다.

식 (2 · 20)의 크기가 매우 작을 때, 즉 식 (2 · 19)의 x가 X에 비해 현저히 작을 때 식 (2 · 19)를 비연성설계로 간주할 수 있다.

위의 정리를 좀 더 자세히 설명한다. 현재의 설계점에서 $FR_0 = ADP_0$라 하자. 현재의 설계점에서 설계파라미터에 ΔDP만큼의 증분이 있다. 기능요구의 증분 ΔFR은 근사적으로 식 (2·18)에서 δ에 Δ를 대입하면 된다. 이때 $(\Delta FR_i)_{\text{diag}} \equiv \frac{\partial FR_i}{\partial DP_i} \Delta DP_i$로 정의하면 $(\Delta FR_i)_{\text{diag}}$는 DP_i가 변화할 때 i번째 대각항에 의해 발생하는 FR_i의 변화량이다. 설계행렬에서 대각항의 크기가 크므로 FR_i의 결정에 DP_i의 영향이 가장 크다. 따라서 $(\Delta FR_i)_{\text{diag}}$이 기능요구의 증분에 가장 많은 부분을 차지한다고 볼 수 있다. 기능요구의 증분에서 $(\Delta FR_i)_{\text{diag}}$을 제외하면 비대각항들에 의한 증분이 된다. 비대각항들에 의한 증분의 영향이 미미할 경우에는 비대각항을 고려하지 않아도 된다. 이를 다시 표현하면

$$\sum_{\substack{j \neq i \\ j=1}}^{n} \left(\frac{\partial FR_i}{\partial DP_j} \right) \Delta DP_j \leq (\Delta FR_i)_{\text{allowable}} \qquad i = 1, \cdots, n \tag{2·21}$$

이 되는데, 여기서 $(\Delta FR_i)_{\text{allowable}}$는 설계자가 허용하는 공차로서 설계자가 정의한다. 식 (2·21)은 비대각항에 의한 영향이 허용공차보다 작도록 하는, 즉 비연성설계로 간주할 수 있는 설계파라미터의 범위를 의미한다.

예제 2·4 비연성설계로 간주할 수 있는 설계파라미터의 범위

다음과 같이 정의된 설계가 있다.

$$\begin{bmatrix} FR_1 \\ FR_2 \end{bmatrix} = \begin{bmatrix} X & x \\ y & X \end{bmatrix} \begin{bmatrix} DP_1 \\ DP_2 \end{bmatrix} \tag{2·22}$$

$X \gg x$, $X \gg y$이며 각 기능요구에 대해 설계자가 정의한 비대각항에 의한 공차 범위를 $(\Delta FR_i)_{\text{allowable}}$, $i = 1, 2$라 할 때

(1) 위의 설계를 비연성설계로 간주할 수 있는 설계파라미터의 범위를 구하여라.

(2) 위의 설계를 비연성화설계로 간주할 수 있는 설계파라미터의 범위를 구하여라.

풀 이

(1) $x\Delta DP_2 \leq (\Delta FR_1)_{\text{allowable}}$와 $y\Delta DP_1 \leq (\Delta FR_2)_{\text{allowable}}$을 모두 만족하는 ΔDP_1과 ΔDP_2의 범위

(2) $x\Delta DP_2 \leq (\Delta FR_1)_{\text{allowable}}$이거나 $y\Delta DP_1 \leq (\Delta FR_2)_{\text{allowable}}$ 중 적어도 하나를 만족하는 ΔDP_1과 ΔDP_2의 범위

2.2.3 물리적 통합

좋은 설계는 단순한 설계라는 속설이 있다. 이 관점에서 볼 때 하나의 DP가 여러 FR을 동시에 만족하면 좋을 것이라는 추측이 나온다. 다시 말해 연성설계가 더 좋은 것이 될 수도 있을 듯이 보인다. 이는 공리적설계를 공부하고 응용함에 있어서 가장 많이 혼동하는 문제 중 하나이다. 그러나 공리적설계의 관점에서 볼 때, 단순한 설계라는 것은 여러 개의 DP가 하나의 물리적 개체를 이루고 있는 물리적 통합(physical integration)으로 설명할 수 있다. 복수의 DP가 복수의 FR을 각각 독립적으로 만족시키는 동시에 그 DP들이 하나의 개체를 이룰 경우, 그 설계의 정보량이 적어질 수 있다. 다음의 예에서 물리적 통합의 의미를 볼 수 있다.

예제 2 · 5 병따개 문제 (NSF, 1998 ; Suh, 1999)

병과 캔을 열 수 있는 기구가 필요하다고 하자. 기능요구는 다음과 같이 정의한다.

FR_1 : 병을 열 수 있는 기구를 설계하여라.

FR_2 : 깡통을 열 수 있는 기구를 설계하여라.

풀 이

그림 2 · 9에서는 병따개와 깡통따개가 하나의 물리적 개체를 이루고 있다. 그러나 기구 양단에 있는 두 개의 DP가 주어진 기능요구들을 독립적으로 만족시킨다. 따라서 그림 2 · 9의 기구는 독립공리를 만족하는 우수한 설계이고, 물리적으로 통합되어 있다. 물론 기능요구나 제한조건에 '두 기능을 동시에 사용할 수 있어야 한다'는 조건이 있다면 다른 설계를 찾아야 한다.

그림 2 · 9 병따개

예제 2 · 6 음료수 깡통 (NSF, 1998 ; Suh, 2001)

우리 주위에서 흔히 접할 수 있는 음료수 깡통이 그림 **2 · 10**에 있다. 이것을 공리적설계 관점에서 분석해 보아라. 음료수 깡통을 제작하는 기술자들에 의하면 여기에는 12개의 FR이 있다고 한다. 자세한 사항은 기밀 사항이므로 발표되지 않았다. 가능한 FR로는 내부의 축방향 원주방향의 압력, 낙하충격에 대한 강도, 쌓았을 때의 강도, 마시기에 편함, 알루미늄 재료사용의 최소화, 표면에 인쇄의 용이성 등이 있을 것이다. 그림 **2 · 10**의 깡통은 12개의 DP로 이루어져 각 FR을 독립적으로 만족하는 한편, 이들이 물리적 통합에 의해 세 개의 요소에 분산되어 통합되어 있다. 세 개의 요소는 몸통, 뚜껑, 마시는 구멍을 여는 요소이다.

설계의 비연성이 유지되면서 물리적으로 통합되어 있을 때는 제품의 복잡성이 감소하게 된다. 관계된 정보량도 줄어들 가능성이 있다. 따라서 물리적 통합은 공리적설계에 어긋나는 것이 아니라 오히려 추천할 사항이다.

그림 2 · 10 음료수 깡통

2.3 정보공리

2.3.1 확률에 의한 정보량 계산

공리적설계에서는 우선 독립공리를 만족하여야 한다. 그런데 독립공리를 만족하는 설계를 두 개 이상 발견할 수 있다. 독립공리를 만족하는 복수의 설계들 중에서 가장 우수한 설계를 선택하려면 정량적으로 정보량이 최소인 것을 선정한다. 그렇다면 정보량은 어떻게 정의할까? 상황에 따라 달라질 수 있지만 정보량은 대체적으로 복잡성(complexity)을 의미한다. 정보공리에서는 정보량을 계산하기 위해 지수를 정의한다. 각 설계의 특성에 따라 정보량의 지수는 다르게 정의할 수 있다. 현재까지 설계에 적용 가능한 지수로 성공의 확률(probability of success)이 알려져 있다. 설계한 제품을 실제로 사용할 때 특정 DP_i가

해당 FR_i를 만족하는 성공의 확률을 p_i라 하자. 이때 정보량 I_i를 다음과 같이 정의한다.

$$I_i = \log_2 \frac{1}{p_i} \tag{2·23}$$

여기서 확률의 역수를 취한 이유는 성공의 확률이 높을수록 정보량을 적게 만들기 위해서이다. 그리고 로그함수를 취한 것은 정보량의 가산성(additivity)을 증진시키기 위함이며, 로그함수의 밑수가 2인 이유는 정보량을 비트(bit) 단위로 표시하기 위함이다.

예를 들어 다음과 같은 비연성설계가 있다고 하자.

$$\begin{bmatrix} FR_1 \\ FR_2 \\ FR_3 \end{bmatrix} = \begin{bmatrix} A_{11} & 0 & 0 \\ 0 & A_{22} & 0 \\ 0 & 0 & A_{33} \end{bmatrix} \begin{bmatrix} DP_1 \\ DP_2 \\ DP_3 \end{bmatrix} \tag{2·24}$$

DP_1, DP_2, DP_3가 각각 FR_1, FR_2, FR_3을 만족할 확률을 p_1, p_2, p_3이라 할 때 총 정보량 I_{total}은 다음과 같다.

$$I_{total} = \sum_{i=1}^{3} I_i = \sum_{i=1}^{3} \log_2\left(\frac{1}{p_i}\right) \tag{2·25}$$

여기서 주의할 점은 정보량은 정의된 기능요구에 관해서만 정의할 수 있다는 것이다.

예제 2·7 비연성설계의 정보량 계산 예

그림 **2·9**의 설계 문제에 대한 정보량을 계산한다. 사용자가 평소의 습관대로 사용할 때, DP_1이 FR_1을 만족할 확률을 0.9라 하고 DP_2가 FR_2를 만족할 확률을 0.85라 가정한다. 이때의 총 정보량은 다음과 같이 계산할 수 있다.

$$\begin{aligned} I_{total} &= I_1 + I_2 = \log_2\left(\frac{1}{0.9}\right) + \log_2\left(\frac{1}{0.85}\right) \\ &= 0.1520 + 0.2345 = 0.3865(\text{bits}) \end{aligned} \tag{2·26}$$

여기서 예제 **2·5**의 문제에서 물리적 통합으로 인해 정보량이 줄어드는 것을 설명해 본다. 통합을 하지 않으면 두 개의 기구를 만들어야 한다. 제작에 필요한 물질의 양을 일정하게 유지하면 각각의 길이나 폭이 줄어들게 된다. 그 결과로 제품의 강도를 유지하기 힘들거나 제품을 손으로 잡기가 불편해져 성공의 확률이 떨어질 것이다. 결과적으로 정보량이 증가하는 셈이다. 따라서 물리적으로 통합된 기구가 정보량이 적은 우수한 설계라는 것을 유추할 수 있다. 그러나 아직도 물리적 통합으로 인한 정보량 감소 정도를 명쾌히 계량화한 연구는 드물다. 이 주제에 대해서는 앞으로 많은 연구가 필요하다.

예제 2 · 8 일정한 길이를 갖는 봉의 제작

성공의 확률을 계산하는 또 다른 방법의 예를 보자. 길이가 1m인 봉을 제작하려 하는데 이에 대한 공차가 ±0.00001m와 ±0.1m의 두 가지 경우가 주어졌다. 각 경우에 대한 정보량을 구하여라.

■ll 풀 이 ll■

두 경우에 동일한 기계를 사용하여 제작한다면 분명 공차가 적을 때 성공의 확률이 작을 것이다. 한편 주어진 길이(호칭길이 혹은 nominal length)가 길수록 공차가 차지하는 비율은 작을 것이다. 따라서 성공의 확률은 다음과 같은 함수가 된다.

$$p = f\left(\frac{\text{tolerance}}{\text{nominal length}}\right) \qquad (2 \cdot 27)$$

식 (2 · 27)을 선형함수로 가정한다면, c를 상수라 할 때 성공의 확률은 다음과 같이 된다.

$$p = c\frac{\text{tolerance}}{\text{nominal length}} \qquad (2 \cdot 28)$$

각 경우에 대한 정보량은 식 (2 · 28)에 주어진 치수를 대입하여 구한다.

비연성화설계에 대한 정보량 계산은 다소 다르다. 독립성이 설계의 순서에 의해 만족되는 이유로 뒷부분의 성공확률은 앞부분의 성공확률에 의존한다. 따라서 조건부확률이 된다. 다음과 같은 비연성화설계가 있다고 하자.

$$\begin{bmatrix} FR_1 \\ FR_2 \end{bmatrix} = \begin{bmatrix} X & 0 \\ X & X \end{bmatrix} \begin{bmatrix} DP_1 \\ DP_2 \end{bmatrix} \qquad (2 \cdot 29)$$

DP_1이 FR_1을 만족할 확률을 p_1이라 하고, DP_1이 FR_1을 만족한 상태에서 DP_2가 FR_2를 만족할 확률은 조건부확률인데 이를 p_{21}이라 하자. 그러면 FR_1, FR_2를 모두 만족할 확률 p는 다음과 같이 된다.

$$p = p_1 p_{21} \qquad (2 \cdot 30)$$

따라서 FR_1과 FR_2가 동시에 만족할 확률 p에 의한 총 정보량은 다음과 같이 된다.

$$\begin{aligned} I &= -\log_2 p = -\log_2 (p_1 p_{21}) \\ &= -\log_2 p_1 - \log_2 p_{21} = I_1 + I_2 \end{aligned} \qquad (2 \cdot 31)$$

여기서 소개한 조건부확률에 의한 정보량 계산방법은 정보공리의 특성을 파악하는 데는 유익하다. 그러나 실제문제에 응용하는 경우는 많지 않다. 그보다는 확률밀도함수를 이용하여 정보량을 산출하는 방법이 더 실용적이다.

2.3.2 확률밀도함수와 정보량

확률밀도함수를 이용하여 정보량을 산출하는 것이 더 편리할 수 있다. 이를 도식화한 것이 그림 2 · 11에 있다. 여기서 사용하는 몇 가지 용어를 다음과 같이 정의한다. 설계영역은 설계하려는 제품성능 목표값의 범위이다. 시스템영역은 제조 후 제품의 실제 성능의 분포이고, 공통영역은 설계영역과 시스템영역의 공통범위이다. 설계영역은 범위로만 정의하며, 시스템영역은 성능의 분포함수로 정의한다. 가령 시스템영역이 균등분포를 한다면, 확률밀도함수를 이용하여 그림 2 · 12와 같이 시스템영역을 나타낼 수 있다. 설계는 공통영역을 증가시키는 방향으로 수행해야 한다. 그림 2 · 12에서 시스템영역은 A_{sr}로서 시스템영역의 확률밀도함수(혹은 면적)를 의미하며 A_{cr}은 공통영역의 면적이다. 따라서 제품의 성능이 설계영역 안에 들어올 성공의 확률과 이에 의한 정보량은 다음과 같이 된다.

$$p_s = \frac{A_{cr}}{A_{sr}} \tag{2 · 32a}$$

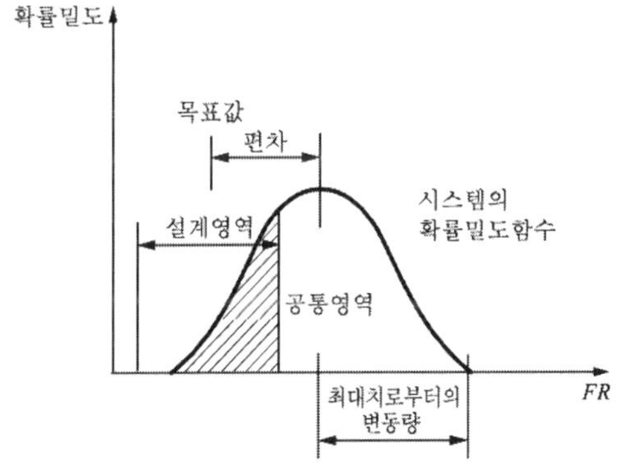

그림 2 · 11 확률밀도함수에 의한 정보량 계산

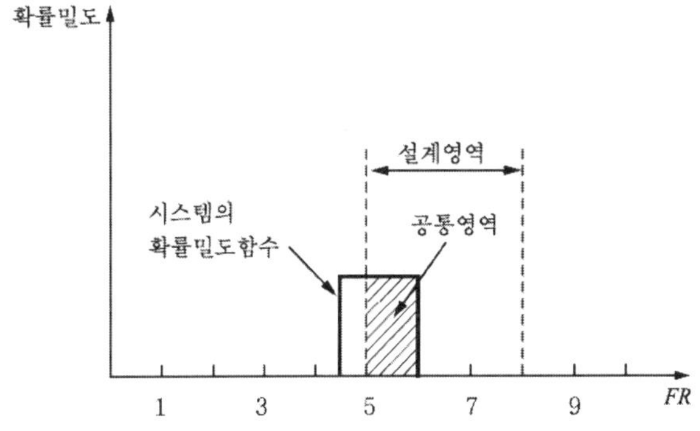

그림 2 · 12 균등분포일 때의 확률분포함수

$$I = -\log_2 \frac{A_{cr}}{A_{sr}} \qquad (2 \cdot 32b)$$

예제 2 · 9 확률밀도함수를 이용한 정보량 계산의 예

이해를 돕기 위해 다소 인위적인 다음의 문제를 생각해 본다. 황광현 씨란 사람이 아파트를 사려고 하는데 다음과 같은 두 가지의 기능요구를 정의한다.

FR_1 : 아파트의 가격이 5000만 원에서 8000만 원 사이에 있게 하여라.

FR_2 : 직장까지 출퇴근 시간이 40분 이내로 되게 하여라.

황광현 씨는 A시와 B시를 고려한다. 각 도시는 표 **2 · 3**의 조건을 가지고 있다. 공리적설계 관점에서 어느 도시에 집을 사는 것이 정보량이 적을까?

표 2 · 3 각 도시별 조건

	A시	B시
가 격	4500~6000만 원	7000~9000만 원
출퇴근 거리	35~50분	20~30분

■ll 풀 이 ll■

표 **2 · 3**의 조건으로부터 시스템영역을 정의한 후 기능요구로부터 설계영역을 결정한다. 모든 확률분포가 균등분포라고 가정할 때, A시의 아파트 가격에 대한 확률분포가 그림 **2 · 12**에 있다. 다른 사항들도 그림 **2 · 12**와 같이 보일 수 있다. A시의 정보량은 다음과 같다.

$$I_{A1} = \log_2\left(\frac{1.5}{1}\right) = 0.59, \quad I_{A2} = \log_2\left(\frac{15}{5}\right) = 1.59 \qquad (2 \cdot 33)$$

$$I_A = I_{A1} + I_{A2} = 2.18(\text{bits}) \qquad (2 \cdot 34)$$

한편 B시의 정보량은 다음과 같이 계산할 수 있다.

$$I_{B1} = \log_2\left(\frac{2}{1}\right) = 1.0, \quad I_{B2} = \log_2\left(\frac{10}{10}\right) = 0.0 \qquad (2 \cdot 35)$$

$$I_B = I_{B1} + I_{B2} = 1.0(\text{bits}) \qquad (2 \cdot 36)$$

아파트를 사기 위한 A시의 정보량 I_A는 2.18이고, B시의 정보량 I_B는 1.0이다. 따라서 정보공리의 관점에서 볼 때 황광현 씨가 사려는 최적의 아파트는 B시에 있다.

식 (2 · 32)를 이용해 정보량을 계산할 때 정보량을 줄일 수 있는 방법을 생각해 본다. 그림 **2 · 11**에서 시스템영역의 평균과 설계영역의 평균간의 차이인 편차(bias)를 먼저 줄이는 것이 효과적이다. 그 후 시스템영역의 분포함수의 표준편차를 줄이면 공통영역이 늘어나고 정보량이 줄어들게 된다. 이는 강건설계와도 관련이 있는데 제7장에서 자세히 설명하도록 한다.

2.3.3 비연성화설계에서의 정보량 계산

비연성설계의 정보량은 식 (2·25)를 사용하여 비교적 쉽게 구할 수 있다. 연성설계에 대한 정보량은 독립공리를 위반하므로 일반적으로 계산하지 않는다. 비연성화설계의 정보량을 구하기 위한 성공의 확률은 앞에서 설명한 조건부확률로 구할 수 있다. 그러나 시스템영역이 확률분포함수로 주어질 경우 하나의 숫자로 확률을 정의해야 하는 조건부확률 계산방법을 그대로 이용하는 방법에는 한계가 있다. 따라서 비연성화설계의 경우에는 일반적으로 정보량을 계산하기 위해 따로 개발된 방법을 사용한다. 이 방법은 공차의 적용방법에 따라 도식적방법과 적분에 의한 방법으로 나눌 수 있다. 도식적 방법은 확률분포함수가 균등분포가 아니거나 세 가지 이상의 기능요구가 있는 경우에는 사용하기 어렵다. 따라서 주로 두 개의 기능요구가 있을 때 사용이 용이하다. 이에 비해 적분에 의한 방법은 다수의 기능요구가 존재할 때에도 정보량 계산이 가능하다. 그러나 중적분을 포함하게 되어 계산이 어려운 단점이 있다. 각 방법을 다음과 같이 설명한다.

설계방정식이 다음과 같은 비연성화설계가 있다고 하자.

$$\begin{bmatrix} FR_1 \\ FR_2 \end{bmatrix} = \begin{bmatrix} X & 0 \\ X & X \end{bmatrix} \begin{bmatrix} DP_1 \\ DP_2 \end{bmatrix} \qquad (2\cdot37)$$

설계파라미터(DP_i)의 임의변동(random variation)에 의한 기능요구(FR_i)의 임의변동은 다음 식과 같다.

$$\begin{bmatrix} \delta FR_1 \\ \delta FR_2 \end{bmatrix} = \begin{bmatrix} \dfrac{\partial FR_1}{\partial DP_1} & \dfrac{\partial FR_1}{\partial DP_2} \\ \dfrac{\partial FR_2}{\partial DP_1} & \dfrac{\partial FR_2}{\partial DP_2} \end{bmatrix} \begin{bmatrix} \delta DP_1 \\ \delta DP_2 \end{bmatrix} = \begin{bmatrix} A_{11} & 0 \\ A_{21} & A_{22} \end{bmatrix} \begin{bmatrix} \delta DP_1 \\ \delta DP_2 \end{bmatrix} \qquad (2\cdot38)$$

설계파라미터의 임의변동이 $\delta\mathrm{DP}$이고 n이 설계파라미터의 수라 할 때, 각 설계파라미터의 제작공차의 범위가 $-\Delta DP_i \le \delta DP_i \le \Delta DP_i$, $i=1, \cdots, n$으로 주어졌다고 하자. 여기서 $0 \le \Delta DP_i$이다. 기능요구의 목표값을 FR^*라 할 때, 설계가 성공했다는 것은 설계파라미터의 임의변동 $\delta\mathrm{DP}$에 대한 기능요구의 임의변동 $\delta\mathrm{FR}$이 설계자가 정의한 공차범위 내에 존재하는 것을 의미한다. 다시 말하면 $-\Delta FR_i \le \delta FR_i \le \Delta FR_i$, $i=1, \cdots, n$을 만족하는 경우이다. 여기에서 $0 \le \Delta FR_i$이다. 이때 목표값이 현재의 설계파라미터 DP^*로 만족했다고 가정한다. 임의변동을 확률변수(random variable)로 취급하여 성공의 확률(p_s)을 다음과 같이 표현할 수 있다.

$$p_s \equiv p(-\Delta FR_1 \le \delta FR_1 \le \Delta FR_1) \cdot p(-\Delta FR_2 \le \delta FR_2 \le \Delta FR_2 \mid -\Delta FR_1 \le \delta FR_1 \le \Delta FR_1) \tag{2 · 39}$$

식 (2 · 38)의 설계방정식의 행렬의 요소인 A_{ij}가 양의 상수(constant)이고, δDP_i가 통계적으로 독립이라고 가정한다. 여기서 δFR_1은 δFR_2에 대하여 통계적으로 종속인 확률변수이다. 식 (2 · 39)의 조건을 DP에 주어진 공차와 식 (2 · 38)을 이용해서 DP영역(domain)에 표현하면 다음과 같다.

$$-\Delta FR_1 \le A_{11}\delta DP_1 \le \Delta FR_1 \tag{2 · 40a}$$

$$-\Delta FR_2 \le A_{21}\delta DP_1 + A_{22}\delta DP_2 \le \Delta FR_2 \tag{2 · 40b}$$

$$-\Delta DP_1 \le \delta DP_1 \le \Delta DP_1 \tag{2 · 40c}$$

$$-\Delta DP_2 \le \delta DP_2 \le \Delta DP_2 \tag{2 · 40d}$$

한편 식 (2 · 40)을 FR영역(domain)으로 사상하면 다음과 같다.

$$-\Delta FR_1 \le \delta FR_1 \le \Delta FR_1 \tag{2 · 41a}$$

$$-\Delta FR_2 \le \delta FR_2 \le \Delta FR_2 \tag{2 · 41b}$$

$$-A_{11}\Delta DP_1 \le \delta FR_1 \le A_{11}\Delta DP_1 \tag{2 · 41c}$$

$$-A_{22}\Delta DP_2 \le \delta FR_2 - \frac{A_{21}}{A_{11}}\delta FR_1 \le A_{22}\Delta DP_2 \tag{2 · 41d}$$

여기서 A_{ij}가 음수이면 부등호가 달라질 수 있다.

식 (2 · 40)에 있는 부등식을 모두 만족하는 δDP의 범위는 식 (2 · 39)의 조건을 만족하는 영역이다. 식 (2 · 41)의 부등식을 모두 만족하는 δFR의 범위 역시 식 (2 · 39)의 조건을 만족한다. 이는 최적화이론에서 제한조건을 모두 만족하는 가용영역(feasible region)과 비슷한 개념이라 하겠다. 식 (2 · 40)이나 식 (2 · 41)의 가용영역에서 확률밀도함수를 구하면 식 (2 · 39)의 확률을 계산할 수 있다.

도식적 방법은 식 (2 · 40)이나 식 (2 · 41)의 유용영역을 2차원의 그래프로 나타내어 가용영역의 면적을 구하게 되므로 기능요구와 설계파라미터의 분포가 균일한 경우에만 사용할 수 있다. 식 (2 · 40)의 각 부등식을 만족하는 범위가 그림 **2 · 13**에 있다. 그림 **2 · 13**에서 성공의 확률과 정보량은 다음의 식으로 계산할 수 있다.

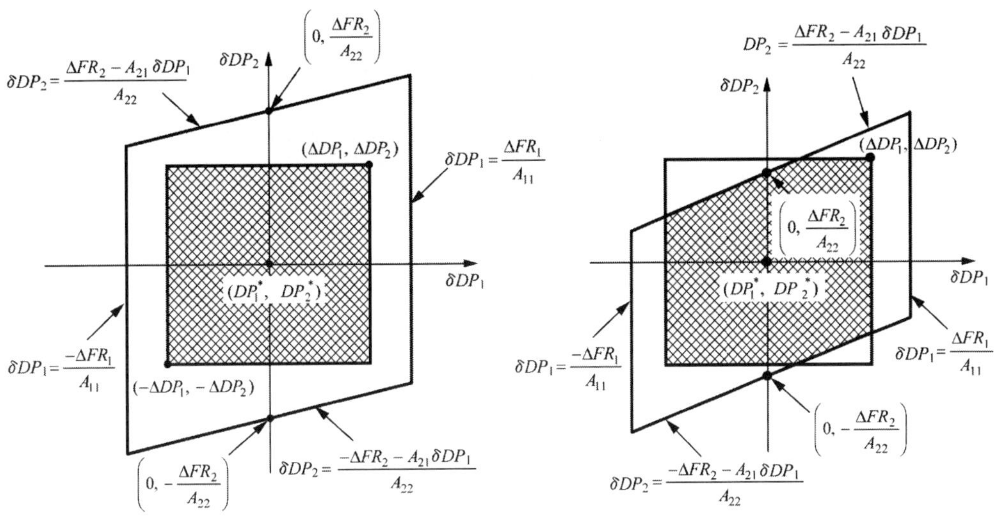

(a) 성공의 확률 = 1 인 경우　　　(b) 성공의 확률 < 1 인 경우

그림 2·13 DP영역에서 비연성화설계의 성공의 확률

$$p_s = \frac{A_{\mathrm{f}}}{A_{\mathrm{dp}}} \tag{2·42a}$$

$$I = -\log_2 p_s \tag{2·42b}$$

여기서 A_{f}는 가용영역으로서 그림 **2·13**에서 빗금 친 부분이다. A_{dp}는 설계파라미터에 주어진 공차에 의한 면적으로서 그림 **2·13**에서 $4\Delta DP_1 \Delta DP_2$이다.

FR영역에서의 도식적인 방법이 그림 **2·14**에 있다. 이는 기능요구의 영역이므로 앞에서 정의한 시스템영역, 설계영역, 공통영역을 정의할 수 있다. 성공의 확률과 정보량은 앞의 식 (2·32)와 같이

$$p_s = \frac{A_{\mathrm{cr}}}{A_{\mathrm{sr}}} \tag{2·43a}$$

$$I = -\log_2 p_s \tag{2·43b}$$

이다. 여기서 A_{sr}은 시스템영역의 면적으로서 그림 **2·14**의 평행사변형의 면적이며, 설계영역은 그림 **2·14**의 직사각형 부분이다. 공통영역 A_{cr}은 시스템영역과 설계영역의 공통부분의 면적인데 이는 식 (2·41)의 가용영역과 동일하다. 여기서 주의할 점은 그림 **2·13** (a)에서는 성공의 확률이 1이나 그림 **2·13** (b)에서는 1이 아니다. 그림 **2·13** (b)의 빗금 친 부분이나 그림 **2·14** (b)의 공통영역의 면적을 구하는 것은 다소 복잡하다. 그림 **2·14** (b)의 경우에 대한 성공의 확률에 대한 결과식은 다음과 같다.

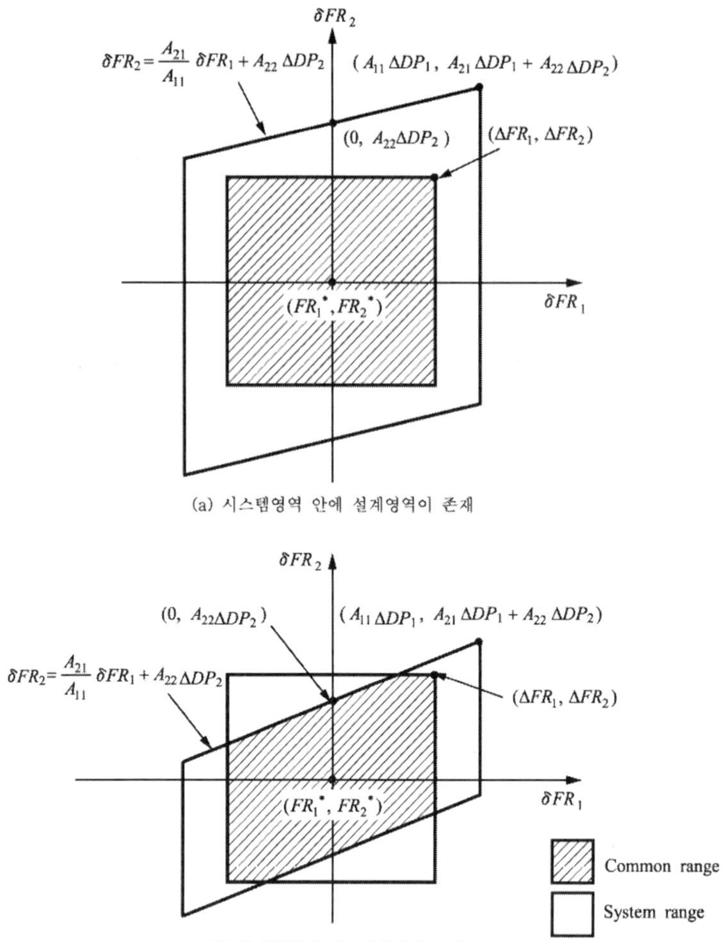

(a) 시스템영역 안에 설계영역이 존재

(b) 설계영역과 시스템영역이 교차

그림 2 · 14 FR영역에서 비연성화설계의 성공의 확률

$$p_s = \frac{\Delta FR_1}{A_{11}\Delta DP_1} - \frac{\frac{A_{11}}{A_{21}}\left(\frac{A_{21}}{A_{11}}\Delta FR_1 + A_{22}\Delta DP_2 - \Delta FR_2\right)^2}{4(A_{11}\Delta DP_1)\cdot(A_{22}\Delta DP_2)} \tag{2 · 44}$$

성공의 확률은 중적분을 이용하여 구할 수도 있다. 이는 DP영역에서 행하는데 앞에서 언급한 바와 같이 그림 **2 · 13**의 가용영역, 즉 빗금 친 부분의 확률밀도함수를 구하면 된다. δDP_1과 δDP_2의 분포함수를 각각 $p_{\delta DP_1}$, $p_{\delta DP_2}$라 하자. 가용영역을 Ω라 할 때 가용영역에서의 확률밀도함수, 즉 성공의 확률은

$$\iint_{\Omega} p_{\delta DP_1} p_{\delta DP_2}\, \mathrm{d}\,\delta DP_2\, \mathrm{d}\,\delta DP_1 \tag{2 · 45}$$

이다.

가용영역이 그림 **2 · 13** (a)와 같을 경우에는 적분이 간단하나 그림 **2 · 13** (b)와 같이 될 경우에는 다소 어려워진다. 이를 해결하기 위해 다음과 같은 단위계단함수 $u(x)$를 이용한다.

$$u(x-x^{*}) = 1 : x \geq x^{*} \text{일 때}$$
$$= 0 : x < x^{*} \text{일 때} \qquad (2 \cdot 46)$$

그림 **2 · 15**가 이러한 단위계단함수를 보여주고 있다. 단위계단함수를 이용하면 가용영역 뿐 아니라 전 영역에서 확률분포를 다음과 같이 정의할 수 있다.

$$p_1 = p_{\delta DP_1} \cdot [u(\delta DP_1 - (-\varDelta DP_1)) - u(\delta DP_1 - \varDelta DP_1)] \qquad (2 \cdot 47a)$$

$$p_2 = p_{\delta DP_2} \cdot [u(\delta DP_2 - (-\varDelta DP_2)) - u(\delta DP_2 - \varDelta DP_2)] \qquad (2 \cdot 47b)$$

이를 가용영역에서 적분하려면 그림 **2 · 13** (a)(b)의 평행사변형을 적분구간으로 하면 된다. 적분구간을 산출해 보면 다음과 같다.

$$\left(\frac{-\varDelta FR_1}{A_{11}} \leq \delta DP_1 \leq \frac{\varDelta FR_1}{A_{11}}, \right.$$
$$\left. \frac{-\varDelta FR_2 - A_{21}\,\delta DP_1}{A_{22}} \leq \delta DP_2 \leq \frac{\varDelta FR_2 - A_{21}\,\delta DP_1}{A_{22}} \right) \qquad (2 \cdot 48)$$

이를 바탕으로 성공의 확률을 구하면, p_1과 p_2는 통계적으로 독립이므로 식 (2 · 47)을 이용하여 다음과 같이 구할 수 있다.

$$\int_{\frac{-\varDelta FR_1}{A_{11}}}^{\frac{\varDelta FR_1}{A_{11}}} \int_{\frac{-\varDelta FR_2 - A_{21}\delta DP_1}{A_{22}}}^{\frac{\varDelta FR_2 - A_{21}\delta DP_1}{A_{22}}} p_1 p_2 \;\mathrm{d}\,\delta DP_2\, \mathrm{d}\,\delta DP_1 \qquad (2 \cdot 49)$$

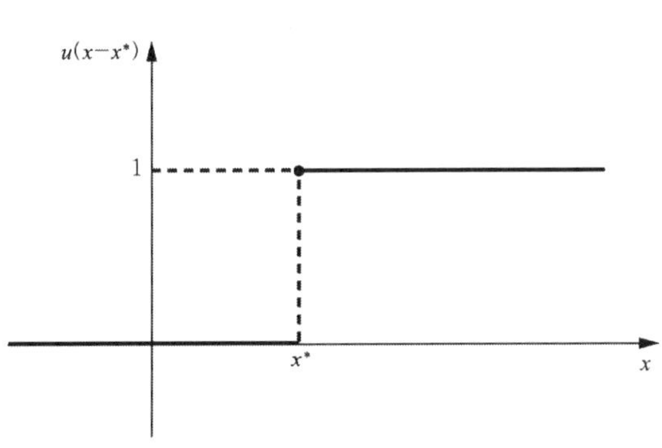

그림 2 · 15 단위계단함수

어떤 경우에 현재의 설계파라미터에 의해 기능요구의 목표값 FR^*를 정확히 만족시키지 못할 수도 있다. 이때는 다음과 같은 관계식이 성립할 것이다.

$$FR_c = ADP_c \tag{2·50a}$$

$$FR_c \neq FR^* \tag{2·50b}$$

여기서 $DP_c = [DP_{c1} \;\; DP_{c2}]^T$, FR_c는 각각 현재의 설계파라미터와 그에 의한 기능요구의 값이다. 도식적방법에 의한 성공의 확률은 그림 **2·13**과 그림 **2·14**의 직사각형과 평행사변형을 DP_c가 원점이 되도록 좌표변환을 통해 옮기면 된다.

이때 식 (2·40)이 조건은 다음과 같다.

$$FR_1^* - \Delta FR_1 \leq A_{11}(DP_{c1} + \delta DP_1) \leq FR_1^* + \Delta FR_1 \tag{2·51a}$$

$$\begin{aligned} FR_2^* - \Delta FR_2 &\leq A_{21}(DP_{c1} + \delta DP_1) + A_{22}(DP_{c2} + \delta DP_2) \\ &\leq FR_2^* + \Delta FR_2 \end{aligned} \tag{2·51b}$$

$$-\Delta DP_1 \leq \delta DP_1 \leq \Delta DP_1 \tag{2·51c}$$

$$-\Delta DP_2 \leq \delta DP_2 \leq \Delta DP_2 \tag{2·51d}$$

확률분포 p_1, p_2는 식 (2·47)을 다시 사용할 수 있다. 식 (2·47)과 식 (2·51)을 이용하여 성공의 확률은 다음과 같이 계산할 수 있다.

$$\int_{\frac{FR_1^* - A_{11}DP_{c1} - \Delta FR_1}{A_{11}}}^{\frac{FR_1^* - A_{11}DP_{c1} + \Delta FR_1}{A_{11}}} \int_{\frac{FR_2^* - A_{21}DP_{c1} - A_{22}DP_{c2} - \Delta FR_2 - A_{21}\delta DP_1}{A_{22}}}^{\frac{FR_2^* - A_{21}DP_{c1} - A_{22}DP_{c2} + \Delta FR_2 - A_{21}\delta DP_1}{A_{22}}} p_1 p_2 \; \mathrm{d}\,\delta DP_2 \, \mathrm{d}\,\delta DP_1 \tag{2·52}$$

적분법에 의한 장점은 다수의 설계파라미터에 대해서도 성공의 확률을 구할 수 있다는 점이다. 이러한 문제에 대해서는 설계파라미터의 수가 n이라 할 때 일반적으로 다음과 같은 중적분식을 사용한다.

$$\int_{\frac{FR_1^* - A_{11}DP_{c1} - \Delta FR_1}{A_{11}}}^{\frac{FR_1^* - A_{11}DP_{c1} + \Delta FR_1}{A_{11}}} \cdots \int_{\frac{FR_n^* - \sum_{i=1}^{n} A_{ni}DP_{c1} - \Delta FR_n - \sum_{i=1}^{n-1} A_{ni}\delta DP_i}{A_{nn}}}^{\frac{FR_n^* - \sum_{i=1}^{n} A_{ni}DP_{c1} + \Delta FR_n - \sum_{i=1}^{n-1} A_{ni}\delta DP_i}{A_{nn}}} p_1 \cdots p_n \, \mathrm{d}\,\delta DP_n \cdots \mathrm{d}\,\delta DP_1 \tag{2·53}$$

적분법은 식 (2·41)의 가용영역에 대해서 FR의 분포함수에 대해 시행할 수도 있다.

이상으로 성공의 확률 및 정보량 계산에 대해 설명하였다. 이는 여러 단계로 이루어진 FR-DP를 가지는 설계에도 적용할 수 있다. 행렬의 곱셈식을 전개하면 모든 하위 단계의

FR-DP로 이루어진 하나의 설계행렬을 만들 수 있다. 이러한 행렬로 정의되는 식에 위의 방법을 적용하면 된다. 한편 연성설계의 정보량 계산을 생각해 볼 수 있다. 위의 방법을 적절히 변형하면 불가능한 것은 아니다. 그러나 그 과정이 대단히 복잡할 뿐만 아니라, 실제로 설계를 수행할 경우 연성설계를 도출하지는 않으므로 여기서는 연성설계의 정보량에 관한 논의는 생략한다.

예제 2·10 비연성화설계의 정보량 계산 1

다음의 비연성화설계가 있다.

$$\begin{bmatrix} FR_1 \\ FR_2 \end{bmatrix} = \begin{bmatrix} 3 & 0 \\ 2 & 5 \end{bmatrix} \begin{bmatrix} DP_1 \\ DP_2 \end{bmatrix} \tag{2·54}$$

$\begin{bmatrix} DP_1^* \\ DP_2^* \end{bmatrix} = \begin{bmatrix} 1 \\ 1 \end{bmatrix}$, $\begin{bmatrix} FR_1^* \\ FR_2^* \end{bmatrix} = \begin{bmatrix} 3 \\ 7 \end{bmatrix}$이며 설계파라미터의 공차는 $\Delta DP_1 = 0.3$, $\Delta DP_2 = 0.3$이고, 기능요구의 허용공차는 $\Delta FR = 1.5$이다. 여기서 δDP_i는 공차구간 내에서 균등분포를 한다.

(1) DP영역에서 도식적 방법에 의한 정보량을 구하여라.

(2) FR영역에서 도식적 방법에 의한 정보량을 구하여라.

(3) DP영역에서 적분법에 의한 정보량을 구하여라.

풀 이

(1) DP영역에서 도식적 방법에 의한 정보량

식 (2·40)에 의해 DP영역에서의 유용영역은 다음과 같다.

$$-1.5 \le 3\delta DP_1 \le 1.5 \tag{2·55a}$$

$$-1.5 \le 2\delta DP_1 + 5\delta DP_2 \le 1.5 \tag{2·55b}$$

$$-0.3 \le \delta DP_1 \le 0.3 \tag{2·55c}$$

$$-0.3 \le \delta DP_2 \le 0.3 \tag{2·55d}$$

이를 도식화하여 보이면 그림 **2·16**과 같다. 성공의 확률과 정보량은 식 (2·42)에 의해

$$p_s = \frac{A_f}{A_{dp}} = \frac{0.6\times0.6 - 0.3\times0.12\times0.5\times2}{0.6\times0.6} = 0.9 \tag{2·56a}$$

$$I = \log_2 \frac{1}{p_s} = 0.152(\text{bits}) \tag{2·56b}$$

이다.

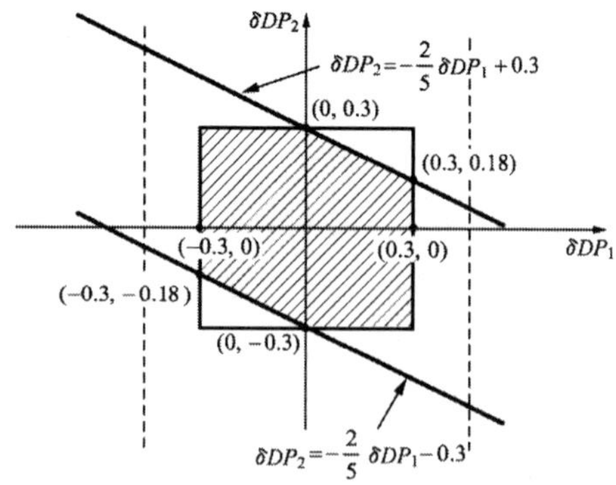

그림 2 · 16 예제 2 · 10 (1)의 DP영역에서의 도식화

(2) FR영역에서 도식적 방법에 의한 정보량

식 (2 · 41)에 의해 FR영역에서 유용영역을 살펴보면 다음과 같다.

$$-1.5 \leq \delta FR_1 \leq 1.5 \tag{2 · 57a}$$

$$-1.5 \leq \delta FR_2 \leq 1.5 \tag{2 · 57b}$$

$$-3\times0.3 \leq \delta FR_1 \leq 3\times0.3 \tag{2 · 57c}$$

$$-5\times0.3 \leq \delta FR_2 - \frac{2}{3}\delta FR_1 \leq 5\times0.3 \tag{2 · 57d}$$

이것을 도식화하면 그림 **2 · 17**과 같다. 성공의 확률과 정보량은 식 (2 · 43)에 의해

$$p_s = \frac{A_{cr}}{A_{sr}} = \frac{1.8\times3-0.9\times0.6\times0.5\times2}{1.8\times3} = 0.9 \tag{2 · 58a}$$

$$I = \log_2\frac{1}{p_s} = 0.152(\text{bits}) \tag{2 · 58b}$$

이다.

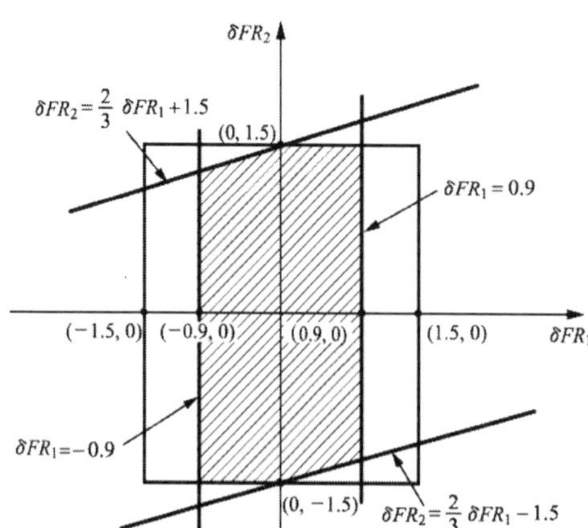

그림 2 · 17 예제 2 · 10 (2)의 FR영역에서의 도식화

(3) DP영역에서 적분법에 의한 정보량

식 (2 · 47)에서

$$p_{\delta DP_1} = p_{\delta DP_2} = \frac{1}{2\varDelta DP_1} = \frac{1}{0.6} \tag{2 · 59a}$$

$$p_1 = 1.667 \times [u(\delta DP_1 - (-0.3)) - u(\delta DP_1 - 0.3)] \tag{2 · 59b}$$

$$p_2 = 1.667 \times [u(\delta DP_2 - (-0.3)) - u(\delta DP_2 - 0.3)] \tag{2 · 59c}$$

성공의 확률과 정보량은 식 (2 · 49)에 의해

$$p_s = \int_{-\frac{1.5}{3}}^{\frac{1.5}{3}} \int_{\frac{-1.5-2\delta DP_1}{5}}^{\frac{1.5-2\delta DP_1}{5}} p_1 p_2 \ \mathrm{d}\delta DP_2 \, \mathrm{d}\delta DP_1 = 0.9 \tag{2 · 60a}$$

$$I = \log_2 \frac{1}{p_s} = 0.152(\text{bits}) \tag{2 · 60b}$$

이다.

예제 2 · 11 비연성화설계의 정보량 계산 2

식 (2 · 54)의 비연성화설계에 대해서 $\begin{bmatrix} DP_{c1} \\ DP_{c2} \end{bmatrix} = \begin{bmatrix} 0.9 \\ 1.1 \end{bmatrix}$일 때 기능요구의 허용공차는 $FR_i^* - \varDelta FR_i \le FR_i \le FR_i^* + \varDelta FR_i$이며, $\begin{bmatrix} FR_1^* \\ FR_2^* \end{bmatrix} = \begin{bmatrix} 3 \\ 7 \end{bmatrix}$, $\varDelta FR_i = 1.5$, $i = 1, 2$이다. 단, 설계파라미터의 공차와 분포는 예제 **2 · 10**과 동일하다.

(1) DP영역에서 도식적 방법에 의한 정보량을 구하여라.

(2) FR영역에서 도식적 방법에 의한 정보량을 구하여라.

(3) DP영역에서 적분법에 의한 정보량을 구하여라.

■ll 풀 이 ll■

(1) DP영역에서 도식적방법에 의한 정보량

식 (2 · 40)을 이 문제에 맞게 변형하면 다음과 같다.

$$FR_1^* - \varDelta FR_1 \le A_{11}(DP_{c1} + \delta DP_1) \le FR_1^* + \varDelta FR_1 \tag{2 · 61a}$$

$$FR_2^* - \varDelta FR_2 \le A_{21}(DP_{c1} + \delta DP_1) + A_{22}(DP_{c2} + \delta DP_2) \le FR_2^* + \varDelta FR_2 \tag{2 · 61b}$$

$$-\varDelta DP_1 \le \delta DP_1 \le \varDelta DP_1 \tag{2 · 61c}$$

$$-\varDelta DP_2 \le \delta DP_2 \le \varDelta DP_2 \tag{2 · 61d}$$

식 (2 · 61)은

$$3 - 1.5 \le 3(0.9 + \delta DP_1) \le 3 + 1.5 \tag{2 · 62a}$$

$$7 - 1.5 \le 2(0.9 + \delta DP_1) + 5(1.1 + \delta DP_2) \le 7 + 1.5 \tag{2 · 62b}$$

$$-0.3 \le \delta DP_1 \le 0.3 \tag{2·62c}$$

$$-0.3 \le \delta DP_2 \le 0.3 \tag{2·62d}$$

이 된다. 이를 도식화하면 그림 **2·18**과 같다. 따라서 성공의 확률과 정보량은

$$p_s = \frac{0.6\times0.6-(0.18\times0.45\times0.5+0.15\times0.06\times0.5)}{0.6\times0.6} = 0.875 \tag{2·63a}$$

$$I = \log_2 \frac{1}{p_s} = 0.193(\text{bits}) \tag{2·63b}$$

이다.

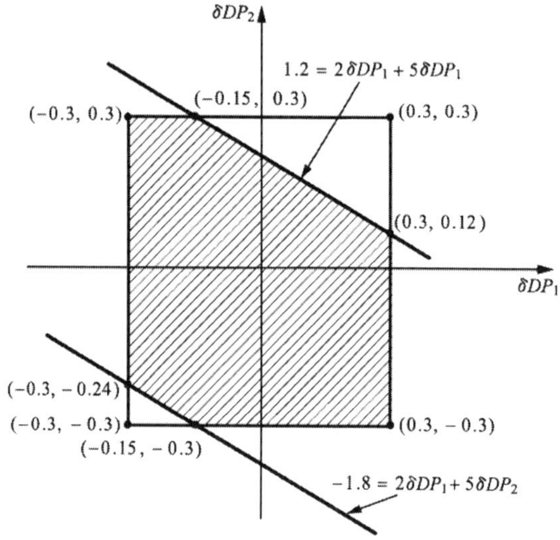

그림 2·18 예제 2·11 (1)의 DP영역에서의 도식화

(2) FR영역에서 도식적 방법에 의한 정보량

식 (2·41)을 이 문제에 맞게 변형하면

$$\begin{bmatrix} FR_{c1} \\ FR_{c2} \end{bmatrix} = \begin{bmatrix} 3 & 0 \\ 2 & 5 \end{bmatrix} \begin{bmatrix} DP_{c1} \\ DP_{c2} \end{bmatrix} = \begin{bmatrix} 2.7 \\ 7.3 \end{bmatrix} \tag{2·64a}$$

$$FR_1^* - \Delta FR_1 \le FR_{c1} + \delta FR_1 \le FR_1^* + \Delta FR_1 \tag{2·64b}$$

$$FR_2^* - \Delta FR_2 \le FR_{c2} + \delta FR_2 \le FR_2^* + \Delta FR_2 \tag{2·64c}$$

$$-A_{11}\Delta DP_1 \le \delta FR_1 \le A_{11}\Delta DP_1 \tag{2·64d}$$

$$-A_{22}\Delta DP_2 \le \delta FR_2 - \frac{A_{21}}{A_{11}}\delta FR_1 \le A_{22}\Delta DP_2 \tag{2·64e}$$

식 (2·64)는

$$-1.2 \le \delta FR_1 \le 1.8 \tag{2·65a}$$

$$-1.8 \leq \delta FR_2 \leq 1.2 \tag{2・65b}$$

$$-0.9 \leq \delta FR_1 \leq 0.9 \tag{2・65c}$$

$$-1.5 \leq \delta FR_2 - \frac{2}{3}\delta FR_1 \leq 1.5 \tag{2・65d}$$

식 (2・65)를 도식화하면 그림 **2・19**와 같다. 이로부터 성공의 확률과 정보량을 계산하면

$$p_s = \frac{A_{cr}}{A_{sr}} = \frac{1.8\times3-(0.45\times0.3\times0.5+1.35\times0.9\times0.5)}{1.8\times3} = 0.875 \tag{2・66a}$$

$$I = \log_2\frac{1}{p_s} = 0.193(\text{bits}) \tag{2・66b}$$

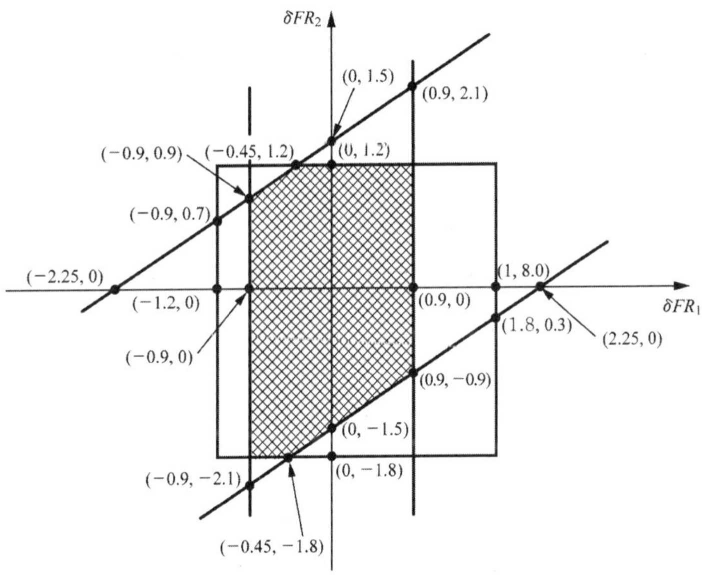

그림 2・19 예제 2・11 (2)의 FR영역에서의 도식화

(3) DP영역에서 적분법에 의한 정보량

중적분 구간은 그림 **2・18**에서 평행사변형의 안이다. 따라서 다음의 중적분으로 성공의 확률을 산출한다.

$$p_s = \int_{-0.4}^{0.6}\int_{\frac{-1.8-2\delta DP_1}{5}}^{\frac{1.2-2\delta DP_1}{5}} p_1 p_2 \, \mathrm{d}\delta DP_2 \, \mathrm{d}\delta DP_1 = 0.875 \tag{2・67}$$

여기서 p_1, p_2는 식 (2・59)와 동일하다.

예제 2 · 12 비연성화설계의 정보량 계산 3

식 (2 · 54)의 비연성화설계에 대해서 설계파라미터의 분포함수 $p_{\delta DP_i}$가 다음과 같다.

$$p_{\delta DP_i} = -\frac{3}{4(\Delta DP_i)^3}(\delta DP_i)^2 + \frac{3}{4(\Delta DP_i)} \qquad i = 1, 2 \tag{2 · 68}$$

(1) 예제 **2 · 10**에서 설계파라미터의 공차는 동일하고 확률분포만 식 (2 · 68)로 대치하여 적분법으로 성공의 확률을 구하여라.

(2) 예제 **2 · 11**에서 설계파라미터의 공차는 동일하고 확률분포만 식 (2 · 68)로 대치하여 적분법으로 성공의 확률을 구하여라.

풀 이

(1) 예제 **2 · 10**에서 식 (2 · 68)의 분포를 가질 때, 적분법에 의한 정보량

i번째 설계파라미터에 대한 확률분포 P_i는 다음과 같다.

$$P_i = \left[-\frac{3}{4\times0.3^3}(\delta DP_i)^2 + \frac{3}{4\times0.3}\right] \times [u(\delta DP_i - (-\Delta DP_i)) - u(\delta DP_i - \Delta DP_i)] \tag{2 · 69}$$

식 (2 · 69)를 식 (2 · 59)에 대입하여 다음의 중적분을 계산하여 성공의 확률과 정보량을 계산한다.

$$p_s = \int_{-\frac{1.5}{3}}^{\frac{1.5}{3}} \int_{\frac{-1.5-2\delta DP_1}{5}}^{\frac{1.5-2\delta DP_1}{5}} p_1 p_2 \ \mathrm{d}\delta DP_2 \, \mathrm{d}\delta DP_1 = 0.978 \tag{2 · 70a}$$

$$I = \log_2 \frac{1}{p_s} = 0.032(\text{bits}) \tag{2 · 70b}$$

(2) 예제 **2 · 11**에서 식 (2 · 68)의 분포를 가질 때, 적분법에 의한 정보량

식 (2 · 69)~(2 · 70)을 그대로 이용하되 중적분 계산을 다음과 같은 구간에서 실시하여 성공의 확률과 정보량을 계산한다.

$$p_s = \int_{-0.4}^{0.6} \int_{\frac{-1.8-2\delta DP_1}{5}}^{\frac{1.2-2\delta DP_1}{5}} p_1 p_2 \ \mathrm{d}\delta DP_2 \, \mathrm{d}\delta DP_1 = 0.953 \tag{2 · 71a}$$

$$I = \log_2 \frac{1}{p_s} = 0.07(\text{bits}) \tag{2 · 71b}$$

이와 같이 확률밀도함수를 이용하여 각종 설계에 대한 정보량을 계산할 수 있는데, 이에 대한 실제적인 예제가 부록 **2.B**에 수록되어 있다.

2.4 설계공리의 응용

대부분의 경우에 독립공리가 만족되면 정보량은 감소한다. 따라서 정보공리가 독립공리에 의존하므로 정보공리는 필요 없는 것으로 간주할 수 있다. 그러나 그렇지 않은 경우가 발생한다. 예를 들어 주어진 기능요구를 만족하는 비연성설계와 연성설계를 발견했다고 하자. 연성설계의 정보량이 비연성설계의 정보량보다 적을 수 있다. 그러면 연성설계가 더 좋은 설계라고 할 수 있는가? 대답은 '그렇지 않다'이다. 이런 경우는 발견한 연성설계보다 정보량이 더 적은 비연성설계나 비연성화설계가 존재함을 암시하는 것이다. 따라서 설계자는 더 좋은 설계를 찾기 위해 노력해야 한다. 다른 예로 비연성설계나 비연성화설계를 복수로 발견할 수도 있다. 독립공리에 대한 만족도가 동일하다고 가정하면, 정보량이 최소가 되는 설계를 최종적으로 선택해야 한다. 설계 시에 두 가지의 공리를 적용하는 흐름도를 그림 2·20과 같이 정의할 수 있다. 부록 2.B에 그림 2·20의 흐름도를 적용한 전형적인 예가 있다.

공리적설계방법을 실제 설계에서 어떤 방식으로 응용할 수 있는가에 대해 살펴본다. 대체로 다음과 같은 분야에 적용할 수 있다고 본다.

(1) 창조적 설계
(2) 기존설계의 분석
(3) 개선설계

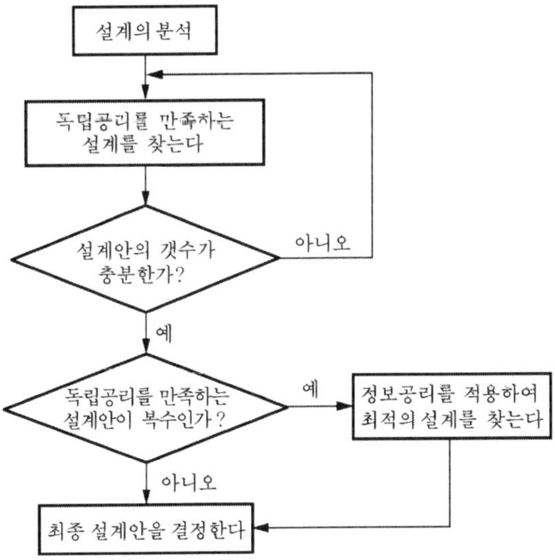

그림 2·20 공리적설계 적용의 흐름도

새로운 기능요구가 정의되고 이러한 기능요구를 만족하는 제품이 없다고 가정하자. 이 경우 새로운 제품을 창출하려고 할 것이다. 창조적 두뇌를 가진 자에 의해 새로운 제품에 대한 아이디어가 창출되는 것이 일반적이다. 그러나 이렇게 창출된 설계안을 구체화하는 방법으로 공리적설계법을 사용할 수 있다. 즉, 아이디어를 공학적으로 분석하고 이를 구현할 수 있는 부품의 선정, 배치 등에 대한 결정 과정을 공리를 이용해서 수행할 수 있다. 그러나 현실적으로 기계설계에 있어서 새로운 창조적 설계를 도출하는 것은 매우 어렵고 드물다. 따라서 기존의 설계를 크게 변화시키고 개선할 수 있다면 창조적 설계로 간주하는 경향이 있다.

이미 생산된 제품들을 평가할 때는 흔히 소비자들의 여론조사를 한다. 그러나 설계자의 관점에서 어떤 제품이 더 우수한가를 평가할 때에 공리적설계방법을 유용하게 도입할 수 있다. 즉, 제품의 목적을 기능요구로 정의하고 현재의 제품을 공리적설계방법으로 분석하여 평가할 수 있다. 이때에는 어떤 제품이 독립공리를 더 만족하는가를 평가하게 된다. 복수의 제품이 동일한 정도로 독립공리를 만족한다면 최소의 정보량을 가지는 가장 좋은 제품을 선정할 수 있다.

끝으로 공리적설계는 현재 제품의 성능을 개선하는 데에 사용할 수 있다. 현재의 제품 성능이 만족스럽지 않을 때나 좀더 좋은 성능을 가지도록 제품 개선을 꾀할 때에 우선 독립공리의 관점에서 기존 제품을 분석한다. 기능요구와 설계파라미터를 정의하고 그들의 관계 정의를 통해 독립공리의 만족여부를 점검해 본다. 독립공리를 만족하지 않을 때는 독립공리를 만족할 수 있도록 개선된 설계를 도출한다. 독립공리를 만족하면 독립공리가 위배되지 않는 범위에서 정보량을 최소화하는 설계파라미터를 도출한다.

예제 2 · 13 **창조적 설계의 예 : 냉장고 설계** (Lee, 1994)

냉장고 설계의 예이다. 일반적인 냉장고의 설계에서 음식물을 오랜 기간 동안 보관하기 위해서는 냉동하고, 짧은 기간 동안 보관하기 위해서는 냉장하는 것을 고려해야 한다. 이로부터 아래와 같은 2개의 기능요구를 정의할 수 있다.

FR_1 : 오랜 기간 음식물의 보관을 위해 냉동하여라.

FR_2 : 짧은 기간 음식물의 보관을 위해 낮은 온도를 유지하여라.

2개의 기능요구를 만족시키기 위하여 2개의 부분으로 구성된 냉장고를 설계할 수 있다. 이에 대한 설계파라미터는 아래와 같다.

DP_1 : 냉동부

DP_2 : 냉장부

첫 번째 수준에서의 설계행렬은 대각행렬로 비연성설계이다. DP_1을 선택함으로써 설계자는 FR_1의 기능을 분해할 수 있다.

FR_{11} : 냉동부의 온도를 -18 ± 2℃로 하여라.

FR_{12} : 냉동부의 온도를 고르게 분포시켜라.

FR_{13} : 냉동부의 상대습도를 50%로 하여라.

마찬가지로 DP_2에 대하여 FR_2의 기능을 분해할 수 있다.

FR_{21} : 냉장부의 온도를 2~3℃로 하여라.

FR_{22} : 냉장부의 온도의 차이를 ±0.5℃ 이내로 분포시켜라.

두 번째 수준의 기능요구를 만족시키는 설계파라미터를 결정해야 한다. 설계파라미터를 결정할 때 기능요구들의 독립성을 만족해야 한다. 주목할 점은 하위수준의 설계파라미터들은 상위수준의 기능요구에 대한 독립성을 유지하는 범위에서 결정해야 한다.

냉동부의 요구사항은 ① 냉동부에 냉각된 공기를 불어넣어 주는 장치, ② 냉동부의 온도를 고르게 하기 위해 냉동 공기를 순환시키는 장치, ③ 순환되는 공기의 온도와 습기를 독립적으로 제어하기 위해 공기상태를 모니터하는 장치로 충족시킬 수 있다. 그러므로 두 번째 수준에서의 설계파라미터는 아래와 같이 선정할 수 있다.

DP_{11} : 냉동부 공기의 온도가 설정 온도보다 높거나 낮을 때 압축기(compressor)를 작동시키는 센서 및 압축기 시스템

DP_{12} : 냉동부에 공기를 불어넣어 주고 일정하게 순환시켜 주는 공기 순환 시스템

DP_{13} : 순환 공기가 이슬점을 넘어서면 습기를 응축시키는 응축기(condenser)

이때의 설계는 식 (2・72)와 같이 비연성화설계이다.

$$\begin{bmatrix} FR_{12} \\ FR_{11} \\ FR_{13} \end{bmatrix} = \begin{bmatrix} X & 0 & 0 \\ X & X & 0 \\ X & 0 & X \end{bmatrix} \begin{bmatrix} DP_{12} \\ DP_{11} \\ DP_{13} \end{bmatrix} \tag{2・72}$$

다음으로 냉장부는 음식물을 보관하기 위하여 냉장실의 온도를 2~3℃로 유지시켜야 한다. 냉장부 또한 냉동부와 비슷하게 압축기를 작동시켜 낮은 온도의 공기를 만들고, 공기를 순환시켜 주는 시스템으로 기능요구를 만족시킬 수 있다. 이에 대한 설계파라미터는 아래와 같다.

DP_{21} : 냉장부 공기의 온도가 설정 온도보다 높거나 낮을 때 압축기(compressor)를 작동시키거나 작동을 중지시키는 센서 및 압축기 시스템

DP_{22} : 냉장부에 공기를 불어넣어 주고 일정하게 순환시켜 주는 공기 순환 시스템

이에 대한 설계행렬은 식 (2・73)과 같이 삼각행렬이다.

$$\begin{bmatrix} FR_{22} \\ FR_{21} \end{bmatrix} = \begin{bmatrix} X & 0 \\ X & X \end{bmatrix} \begin{bmatrix} DP_{22} \\ DP_{21} \end{bmatrix} \tag{2・73}$$

두 번째 수준까지 분해한 설계에 대한 전체 설계방정식은 식 (2・74)와 같이 비연성화설계임을 알 수 있다.

$$\begin{bmatrix} FR_{12} \\ FR_{11} \\ FR_{13} \\ FR_{22} \\ FR_{21} \end{bmatrix} = \begin{bmatrix} X & 0 & 0 & 0 & 0 \\ X & X & 0 & 0 & 0 \\ X & 0 & X & 0 & 0 \\ 0 & 0 & 0 & X & 0 \\ 0 & 0 & 0 & X & X \end{bmatrix} \begin{bmatrix} DP_{12} \\ DP_{11} \\ DP_{13} \\ DP_{22} \\ DP_{21} \end{bmatrix} \qquad (2 \cdot 74)$$

식 (2 · 74)의 전체 설계행렬을 보면, 분해 과정을 통하여 나온 기능요구들이 상위수준에서의 기능요구의 독립성을 유지시키는 것을 알 수 있다.

위의 설계행렬에서 하나의 압축기와 두 개의 팬(fan)을 사용하여 기능요구들을 만족시킬 수 있다. 이때 DP_{11}과 DP_{21}은 두 개의 센서 중 하나가 신호를 감지하면 압축기를 작동시키는 센서/압축기 시스템이다. 그러나 DP_{12}와 DP_{22}의 팬은 각 부의 온도가 설정온도를 벗어나지 않는다면 작동하지 않을 것이다. 이와 같이 하나의 압축기와 두 개의 팬을 가지고 기능요구와 기능요구의 독립성을 만족하는 설계를 구성할 수 있다. 그에 대한 예를 그림 **2 · 21**에 보여주고 있다. 이 외에 독립공리를 만족하는 다양한 설계안을 제시할 수도 있다. 제시된 설계안 중 독립공리를 만족하면서 더 넓은 영역의 온도 제어와 습도 제어가 가능한 해를 최종안으로 선택하면 된다. 이제부터 기존의 냉장고와 새로이 제시된 설계안을 비교해 본다.

전통적인 냉장고는 하나의 압축기와 하나의 팬으로 구성되어 있다. 그림 **2 · 22**와 같이 두 부분(냉동부와 냉장부) 사이의 댐퍼를 조절하여 냉장부를 냉각시킨다. 그러므로 냉장부의 온도를 냉동부와 독립적으로 제어할 수 없다. 냉장부의 온도가 3℃보다 올라가면 댐퍼가 열리지만 냉동부의 압축기와 팬이 작동하지 않으면 FR_{21}을 만족시키지 못하므로 연성설계이다.

부록 **2.A**의 추론 **2.A · 3**에 따르면, 만일 하나의 팬만을 사용하여 기능요구를 독립적으로 만족시킬 수 있다면, 두 개의 팬과 적절한 연결관을 사용하는 것은 최상의 설계 해가 아닐 수도 있다. 독립공리를 만족하는 또 다른 설계안이 존재한다면 두 가지 설계안 중에서 더 좋은 설계를 선택하기 위해서 정보공리를 이용해야 한다.

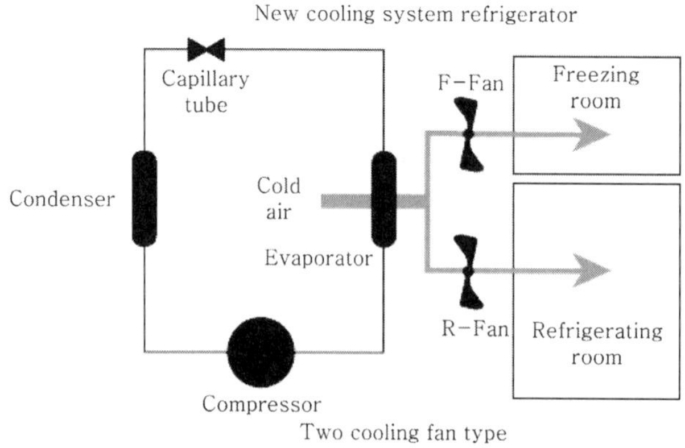

그림 2 · 21 독립공리를 만족하는 새로운 냉장고

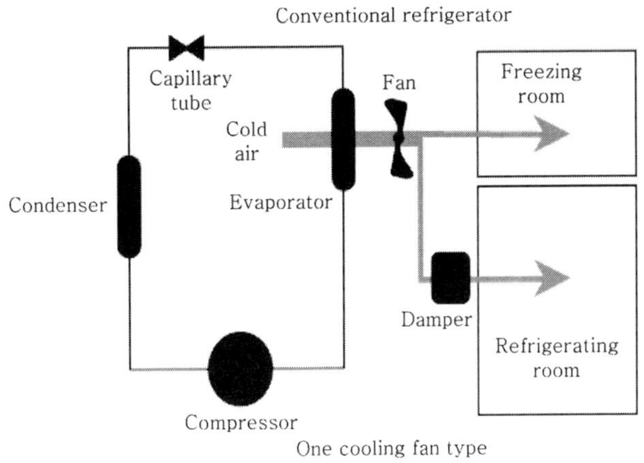

그림 2 · 22 기존의 냉장고

예제 2 · 14 기존설계의 분석에 대한 예 : 액정디스플레이의 고정대 설계(NSF 1998 ; Suh 2000)

액정디스플레이(liquid crystal display, LCD)는 컬러텔레비전 신호의 적색, 녹색, 청색을 투사하는 3개의 LCD 패널(panel)로 구성된 영사장치이다. LCD 영사기의 구성은 그림 **2 · 23**에 있다. LCD 영사 시스템이 정확한 색상을 표시하기 위해서는 청색 이미지에 대하여 패넌의 픽셀(pixel)이 허용 공차 범위 내에 정렬되어야 한다.

픽셀을 정렬시키기 위해서 LCD 영사기는 각 LCD 패널의 회전 및 병신운동을 조절할 수 있는 기구장치를 사용한다. 먼저 3개의 패널 중에서 하나의 픽셀을 기준으로 정한 후, 남은 2개 패널의 픽셀을 적절히 정렬한다. 각 LCD 패널은 LCD 고정대(holder)라고 불리는 조절 기구장치에 장착된다. 픽셀 정렬을 위해서는 적어도 기준이 되는 패널을 제외한 2개의 LCD 고정대가 3 자유도(X축 및 Y축상의 이동, Z축을 중심으로 회전)를 가져야 한다. 이제 기존의 두 가지 LCD 고정대의 설계를 비교하고자 하는데, 이들은 산요(Sanyo)사와 샤프(Sharp)사에서 제작된 것이나.

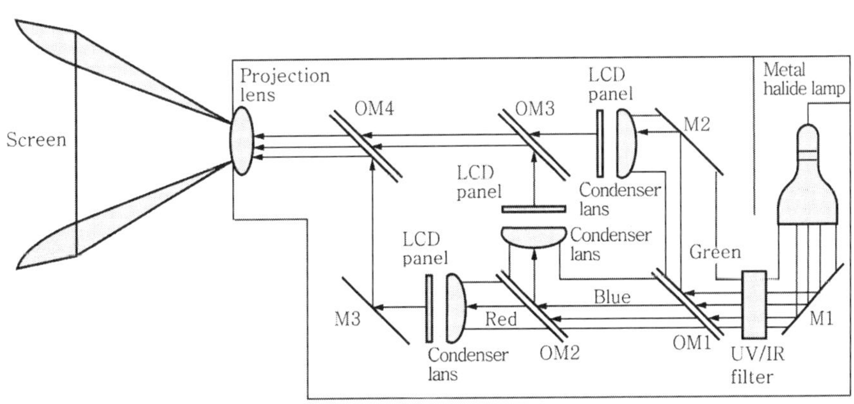

그림 2 · 23 도식화한 LCD

독립공리를 바탕으로 기존설계 중에서 어떤 설계가 3개의 LCD 패널의 픽셀을 정렬시키는 데 우수한지를 판단해 보아라.

풀 이

최상위수준에서의 기능요구와 설계파라미터는 아래와 같다.

FR : LCD 패널의 픽셀을 정렬하여라.

DP : LCD 패널의 픽셀을 정렬할 수 있는 LCD 고정대

각 LCD 패널의 픽셀을 정렬하기 위해서 기능요구를 분해할 수 있다.

FR_1 : X축을 따라서 이동하여라. $T(X)$

FR_2 : Y축을 따라서 이동하여라. $T(Y)$

FR_3 : Z축에 대해서 회전하여라. $R(Z)$

LCD 영사기는 3개의 패널을 사용하는데 그중 하나는 기준위치로 이용된다. 그러므로 픽셀을 정렬시키기 위해서는 3 자유도의 고정대 2개 이상이 필요하다. 지금부터 선정된 상용 LCD 영사기의 설계를 평가하도록 하자.

산요사의 고정대

그림 **2 · 24**는 산요사의 LCD 고정대를 보여주는 그림이다. 산요사의 제품은 3개의 기구로 구성되어 있는데, 모두 리드나사(lead screw)로 고정된다. 나사가 회전하면 해당하는 판이 움직인다. 이와 같이 나사의 회전운동을 판의 이동이나 회전운동으로 변환시킬 수 있다. A-판은 구조물의 측면에 고정되어 있고, 나머지 판들은 A-판 위에서 움직인다. LCD 패널은 C-판에 고정되어 있다. 이들 3개의 기구는 설계파라미터가 된다.

DP_1 : B-판과 나사 1을 결합하는 리드나사

DP_2 : C-판과 나사 2를 결합하는 리드나사

DP_3 : B-판과 나사 3을 결합하는 리드나사

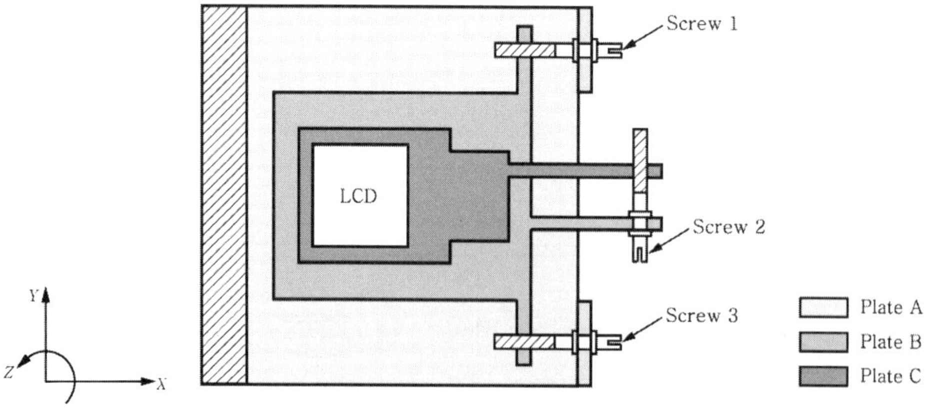

그림 2 · 24 산요사의 LCD판 조정기구

LCD 패널의 중심은 회전 중심과 떨어져 있기 때문에, B-판을 회전시켜 $T(X)$와 $R(Z)$축으로 LCD 패널을 이동 및 회전시킬 수 있다. B-판을 이동시켜 LCD 패널을 이동할 수 있으므로 나사 1을 회전시키면 LCD 패널이 $T(X)$와 $R(Z)$축 방향으로 이동한다. 따라서, DP_1은 FR_1과 FR_3에 영향을 미친다. DP_2는 나사 2와 C-판으로 구성되어 있다. 나사 2를 회전시키면 C-판과 LCD 패널이 B-판을 향하여 Y축으로 움직인다. 나사 2의 회전이 LCD 패널을 $T(Y)$축 방향으로 이동시키므로 DP_2가 FR_2에만 영향을 미친다. DP_3은 DP_1과 같은 구조를 가지고 있다. 나사 3의 회전이 LCD 패널을 패널이 $T(X)$와 $R(Z)$축 방향으로 이동시키므로 DP_3은 FR_1과 FR_3에 영향을 미친다. 이에 대한 설계방정식은 식 (2・75)와 같다.

$$\begin{bmatrix} T(X) \\ T(Y) \\ R(Z) \end{bmatrix} = \begin{bmatrix} X & 0 & X \\ 0 & X & 0 \\ X & 0 & X \end{bmatrix} \begin{bmatrix} DP_1 \\ DP_2 \\ DP_3 \end{bmatrix} \tag{2・75}$$

식 (2・75)의 설계행렬은 연성설계이므로 독립공리를 위배한다.

위의 제품은 픽셀을 정렬하기 위해서 몇 번의 반복 과정을 거쳐야 한다. 예를 들어, 각도가 잘못 조정되면 DP_1을 다시 정렬해야 한다. 그러나 DP_1이 LCD 패널의 $T(X)$축 방향 위치를 변화시키므로 또 다른 오차를 낳는다. 단지 FR_1에만 영향을 미치는 DP가 없으므로 몇 번이고 반복 과정을 거쳐야 한다. 이런 종류의 기구장치는 한 부품에 오류가 생기면 사용자의 오랜 보정이 필요하다.

샤프사의 고정대

샤프사의 제품도 그림 **2・25**와 같이 3개의 기구장치가 있다. 하나는 단순한 리드나사(lead screw)이고 나머지는 더 정밀한 리드나사이다. 각 리드나사에는 경로대(guide way)와 보스(boss)가 장착된다. A-판은 구조물의 측면이 고정되어 있다. LCD패널은 C-판에 장착되어 있어서 같이 움직인다. 이 세 개의 기구들은 설계파라미터로 아래와 같이 정의할 수 있다.

DP_1 : B-판과 나사 1을 결합하는 리드나사

DP_2 : C-판, D-판, 보스 1, 나사 2를 결합하는 리드나사

DP_3 : B-판, E-판, 보스 2, 나사 3을 결합하는 리드나사

나사 1이 회전하면 B-판의 측면이 구조물에 고정된 A-판을 향해 X 방향으로 이동한다. LCD패널은 C-판 위에 장착되었으므로 B-판과 함께 움직이며 X 방향을 따라 움직인다. 이처럼 나사 1과 결합된 DP_1은 FR_1에만 영향을 준다. 나사 2가 회전하면 D-판의 지지경로 때문에 C-판 위에 리벳 처리된 보스 1은 밀려난다. 결과적으로 B-판의 수직홈이 보스 1을 움직이는 경로가 되므로 보스 1은 Y 방향으로 움직인다. 또한 나사 2의 회전으로 LCD패널이 Y 방향으로 이동하므로 DP_2는 FR_2에만 영향을 준다.

나사 3은 E-판을 X 방향으로 이동시키고, E-판의 경로대 때문에 C-판에 리벳 처리된 보스 2를 밀어낸다. 보스 1에는 제한된 방향이 없으므로 C-판은 D-판의 경로대 안쪽에서 보스 2에 대하여 회전을 한다. LCD패널의 중심과 C-판의 회전중심이 떨어져 있으므로 LCD 패널의 회전으로 X 방향과 Y 방향의 이동을 제어할 수 있다. 그러므로 나사 3의 회전으로 LCD패널이 X 방향으로 이동하고 Z축에 대하여 회전한다. DP_3는 FR_1과 FR_3에 영향을 준다. 따라서 설계방정식은 식 (2・76)으로 나타낼 수 있다.

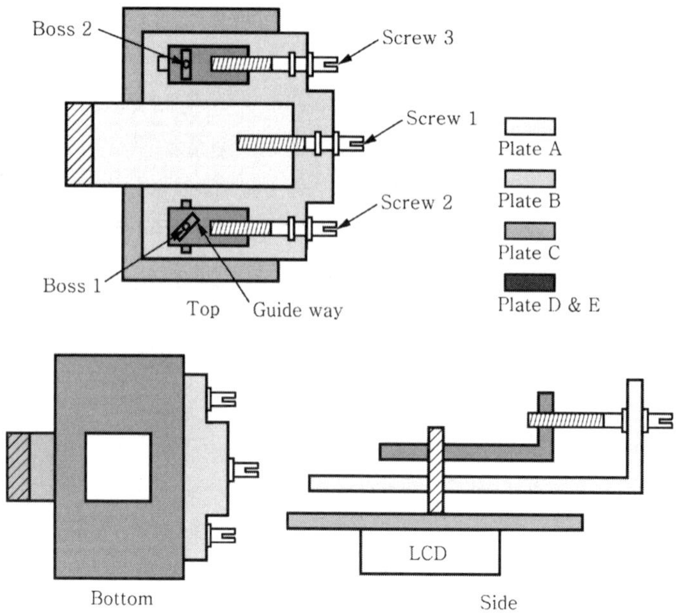

그림 2 · 25 샤프사의 LCD판 조정기구

$$\begin{bmatrix} T(X) \\ T(Y) \\ R(Z) \end{bmatrix} = \begin{bmatrix} A_{11} & 0 & A_{13} \\ 0 & A_{22} & 0 \\ 0 & 0 & A_{33} \end{bmatrix} \begin{bmatrix} DP_1 \\ DP_2 \\ DP_3 \end{bmatrix} \tag{2 · 76}$$

설계행렬이 삼각행렬이므로 이 설계는 비연성화설계이다. 이 설계행렬이 제시하는 순서대로 설계를 진행하면 독립공리를 만족한다. DP_3에 의하여 $R(Z)$축 방향으로 LCD를 정렬시키고 난 후, DP_1과 DP_2로 다른 방향을 정렬시키는 식으로 진행되어야 한다.

위에서 제시하는 방식으로 독립공리를 이용하여 기존설계를 분석하거나 비교할 수 있다. 이 경우에는 하나가 독립공리를 만족하지 않으므로 판정이 간단하다. 그러나 비교 대상의 설계가 모두 독립공리를 만족한다면 R이나 S(독립성의 정도)값을 산정하여 비교하거나 정보량으로 비교할 수도 있다.

예제 2 · 15 개선설계의 예 : 자동변속기의 주차모드 설계 (NSF, 1998 ; Suh, 2000)

자동차의 자동변속기는 주차모드를 가지고 있다. 주차모드는 구동 메커니즘을 잠금으로써 사람이 없을 때 자동차가 움직이는 것을 막아 주도록 설계되어 있다. 자동차가 언덕에 주차되어 있을 때 소비자들은 주차모드에서 기어를 푸는 것이 어렵다고 불평한다. 또한 언덕에서 기어의 주차모드가 풀릴 때 과도한 진동이 발생할 수 있다. 현재의 설계를 그림 **2 · 26**에 나타내었다. 현재의 주차 방법이 어떻게 작동하는지 생각해 보고 더 좋은 설계를 개발하여라.

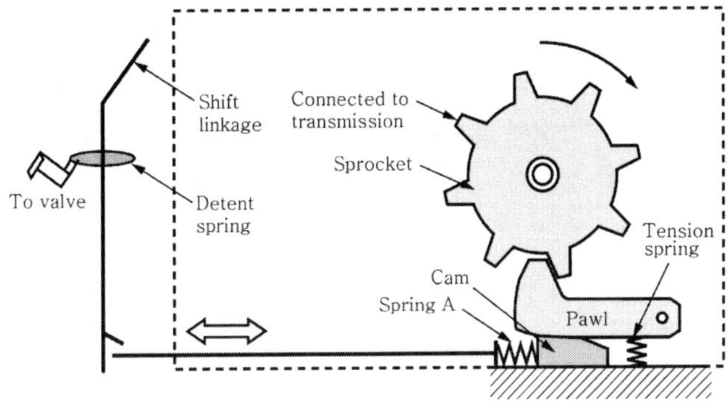

그림 2 · 26 자동변속기를 위한 주차기구의 도식도

풀 이

현재의 설계를 개선하기 위해 현재 설계의 작동방식을 알아야 한다. 그림 **2 · 26**에서는 멈춤쇠(pawl)가 모드전환장치(shift linkage)에 의해 스프라켓(sprocket)에 잠겨서 주차모드 위치에 있음을 알 수 있다. 스프라켓은 변속기에 부착되어 있다. 주차모드 위치로 모드전환 장치가 움직이면 멈춤스프링(detent spring)이 유압시스템을 작동시킨다. 이때 캠에 부착된 스프링 A를 밀게 된다. 모드전환 동작이 발생하면 캠 표면형상이 멈춤쇠를 물림위치로 밀어 내고 그림 **2 · 26**과 같이 캠이 내부로 들어가서 스프라켓을 고정위치에 놓고 잠그도록 되어 있다. 이렇게 해서 자동차는 주차모드에 있게 된다.

자동차가 움직이는 동안에는 멈춤쇠가 스프라켓과 맞물릴 수 없다. 왜냐하면 자동차가 시속 4.8km 이상으로 움직일 때 스프라켓과 멈춤쇠 사이에 충격하중이 작용하기 때문이다. 스프라켓의 치형과 멈춤쇠의 치형은 그들 사이의 충격하중이 스프링 A에 의해 발생되는 하중보다 클 때 물릴 수 없도록 되어 있다. 충격하중이 스프링하중보다 더 크면 멈춤쇠를 뒤로 치게 되어 캠이 주차모드 위치로 이동하지 못하도록 예방하고 있다.

자동차가 언덕 위에 주차되어 있을 때, 자동차의 무게가 스프라켓에 토크를 발생시키고 토크는 치형과 멈춤쇠를 통해 캠에 전달된다. 따라서 자동변속기가 모드전환장치를 움직여 주차모드에서 벗어나려면, 캠과 멈춤쇠 사이의 마찰력보다 큰 힘으로 캠을 당겨야 한다. 캠을 주차모드 위치의 바깥으로 끌어당기면, 멈춤쇠가 스프라켓과 물리는 위치의 바깥으로 당겨지고 인장스프링에 의해 멈춤쇠가 내려온다.

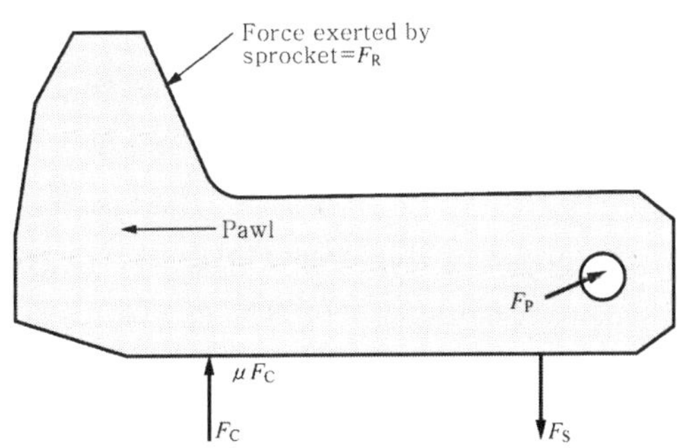

그림 2 · 27 멈춤쇠에 작동하는 힘의 자유물체도

그림 **2 · 27**은 멈춤쇠에 작용하는 힘을 보여주는 자유물체도이다. F_R은 멈춤쇠와 스프라켓 사이에 작용하는 반력이다. F_C는 멈춤쇠와 캠 사이의 반력, F_S는 인장스프링의 스프링력, F_P는 핀에 의해 멈춤쇠에 작용하는 반력이다. 그리고 μ는 멈춤쇠와 캠 사이의 마찰계수이다. 멈춤쇠 치형의 기울기가 증가함에 따라 F_R이 증가하면 반력 F_C와 마찰력 μF_C가 증가한다. 스프링력 F_S는 캠이 제거될 때까지는 일정하나 스프링의 인장에 따라 변한다.

여기서의 기능요구는 다음과 같이 정의할 수 있다.

FR_1 : 고정된 위치에서 멈춤쇠를 물릴 수 있게 하여라.

FR_2 : 고정된 위치로부터 멈춤쇠를 풀 수 있도록 하여라.

FR_3 : 우연한 물림을 방지하여라.

FR_4 : 멈춤쇠가 물리는 위치를 유지할 수 있도록 하여라.

FR_5 : 차량에 의해 전달되는 하중을 다시 전달하여라.

현재 설계는 위와 같은 기능요구를 만족시키기 위해서 아래와 같은 설계파라미터를 갖는다.

DP_1 : 캠 형상의 경사진 면

DP_2 : 인장 스프링

DP_3 : 스프라켓의 치형과 멈춤쇠의 치형 / 스프링 A / 모드전환장치 / 인장스프링

DP_4 : 캠 형상의 평평한 표면

DP_5 : 멈춤쇠 / 캠의 평평한 표면

설계방정식은 식 (2 · 77)과 같다.

$$\begin{bmatrix} FR_1 \\ FR_2 \\ FR_3 \\ FR_4 \\ FR_5 \end{bmatrix} = \begin{bmatrix} X & 0 & 0 & X & X \\ X & X & 0 & 0 & X \\ 0 & X & X & 0 & 0 \\ 0 & X & 0 & X & X \\ X & X & 0 & X & X \end{bmatrix} \begin{bmatrix} DP_1 \\ DP_2 \\ DP_3 \\ DP_4 \\ DP_5 \end{bmatrix} \tag{2 · 77}$$

따라서 현재 설계는 연성설계이다. 이를 자세히 보면 변속기에 의해 전달되는 차량의 하중을 멈춤쇠와 캠이 지탱하기 때문이다. 언덕의 경사가 증가함에 따라 캠에서의 법선력과 마찰력이 증가하므로 주차모드를 해제하는 것이 더욱 어려워진다.

새롭게 제안된 설계

그림 **2 · 28**은 새롭게 제안된 설계를 나타낸다. 새로운 설계는 이전의 설계와 다른 치형을 가진 스프라켓을 포함하고 있다. 치형의 외곽 모서리 근처의 경사진 면은 멈춤쇠가 우연히 물리는 것을 방지하기 위한 것이다. 그리고 멈춤쇠 치형 윗부분의 평평한 면은 차량의 하중을 전달한다. 멈춤쇠의 모서리는 차량이 시속 4.8km로 움직일 때 우연한 물림을 방지하기 위해서 경사져 있다. 따라서 그림 **2 · 28**의 F_R과 F_P가 동일선상에 놓이게 되어 캠과 멈춤쇠 사이의 힘이 제거되는 효과가 있다. 그리고 차량 하중은 캠이 아닌 핀에 의해서 전달된다. 새로운 설계에서 캠은 차량 하중을 전달하지 않는다. F_P는 대략 F_R과 같다.

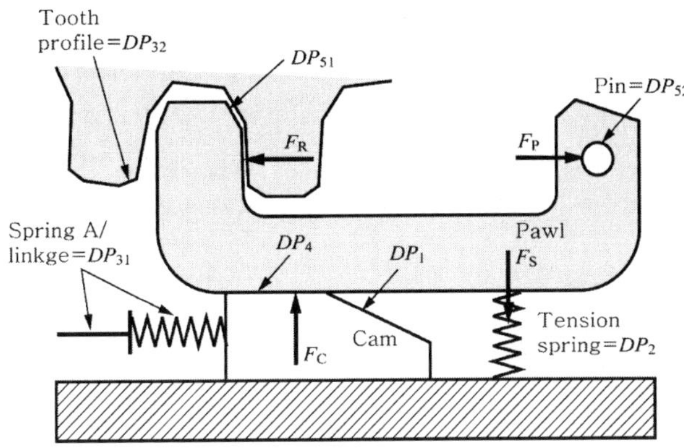

그림 2·28 새롭게 제안된 설계

기능요구는 이전과 같으며 설계파라미터는 그림 **2·28**에 보여지고 있다.

DP_1 : 캠 형상의 경사진 면

DP_2 : 인장 스프링

DP_3 : 스프라켓 휠과 멈춤쇠의 치형/스프링 A/모드전환장치

DP_4 : 캠형상의 평평한 면

DP_5 : 멈춤쇠의 평평한 면/스프라켓과 핀

설계방정식은 다음과 같다.

$$\begin{bmatrix} FR_4 \\ FR_2 \\ FR_3 \\ FR_1 \\ FR_5 \end{bmatrix} = \begin{bmatrix} X & 0 & 0 & 0 & 0 \\ X & X & 0 & 0 & 0 \\ 0 & X & X & 0 & 0 \\ X & 0 & 0 & X & 0 \\ 0 & 0 & 0 & 0 & X \end{bmatrix} \begin{bmatrix} DP_4 \\ DP_2 \\ DP_3 \\ DP_1 \\ DP_5 \end{bmatrix} \tag{2·78}$$

이것은 비연성화된 설계이다.

DP_3과 DP_5의 특성으로부터 다음과 같이 FR_3과 FR_5를 분해할 수 있다.

FR_{31} : 스프라켓으로 멈춤쇠를 밀어 넣는 힘을 제어하여라.

FR_{32} : 스프라켓의 회전 시 반력을 발생시켜라.

이에 상응하는 DP는 다음과 같다.

DP_{31} : 스프링 A

DP_{32} : 스프라켓 치형 형상

이에 관계된 설계행렬은 삼각행렬이다.

FR_5와 DP_5는 다음과 같이 분해할 수 있다.

FR_{51} : 스프라켓의 힘을 멈춤쇠로 전달하여라.

FR_{52} : 전달된 하중을 다시 전달하여라.

DP_{51} : 멈춤쇠의 수직에 가까운 면과 스프라켓 치형 형상

DP_{52} : 멈춤쇠의 수직면에 작용하는 힘 벡터와 동일선상에 위치한 핀

캠과 멈춤쇠 사이의 반력을 최소화하기 위해서는 스프라켓과 멈춤쇠 사이의 반력이 수평에 가까워야 한다. DP_{51}은 이를 위해 수직에 가까운 면이어야 한다. 멈춤쇠와 스프라켓 치형 사이의 작은 경사도는 멈춤쇠의 후퇴량을 적게 하기 위해 이용될 것이다. 여기에서의 설계 행렬은 대각행렬이다. 이상으로 독립공리를 만족하는 개선된 설계를 발견하였다. 물론 독립공리를 적용할 때 더 좋은 아이디어로 더 좋은 개선설계의 결과를 얻을 수도 있을 것이다.

2.5 설계공리를 응용한 소프트웨어 설계

2.5.1 소프트웨어 설계

최근 들어 소프트웨어의 중요성은 공학 전 분야에 걸쳐 널리 인식되고 있다. 소프트웨어는 컴퓨터를 운용하기 위한 기술이나 방법론을 말한다. 그리고 소프트웨어 공학(software engineering)은 최소의 경비로 신뢰도 높은 소프트웨어를 생산하기 위한 기법과 도구를 의미한다. 소프트웨어의 개발에 필요한 이론적인 토대는 소프트웨어 공학에서 찾아볼 수 있다. 그러나 공학 분야의 응용 소프트웨어를 개발하는 개발자들은 소프트웨어 공학에 대해 전반적인 이해가 부족하다. 소프트웨어 공학을 설명하는 용어의 추상성 등으로 인해 개발자 나름의 방식대로 소프트웨어를 개발하려는 경향이 있다. 이러한 방식은 비체계적이며 경험에 크게 의존할 뿐 아니라 개발 과정의 문서화가 충분히 이루어지지 않는다. 따라서 소프트웨어 개발 이후에 나타나는 유지보수(maintenance)나 개정(modification) 및 확장(extension) 등에 많은 시간과 노력이 필요하다.

소프트웨어 공학에서는 이러한 문제점들을 두 가지 접근방법으로 해결하려 한다. 첫째는 초기에 시간과 노력을 집중 투자하여 독립적인 모듈(module)을 설계하고 이들 모듈들을 이용하여 소프트웨어를 설계함으로써 향후에 필요한 노력을 줄이려 한다. 둘째는 객체지향(object oriented)언어 등의 보다 진보된 컴퓨터 언어를 사용함으로써 개발자들의 부담을 덜어 보려는 노력이다. 그러나 이러한 해결 방법을 적극적으로 사용하는 소프트웨어 개발자라 하더라도, 소프트웨어 생명주기(life cycle) 동안에 발생되는 오류수정(debugging), 유지보수 및 확장 등에서 나타나는 소프트웨어의 고전적인 문제점들을 완전히 해결하지는 못하고 있다. 이러한 문제점의 가장 근본적인 이유는 각각의 독립적인 모듈들이 소프트웨어의 실행 과정에 따라 서로 기능적으로 연성(couple) 관계를 이루고 있기 때문이다.

앞에서 설명하였듯이 공리적설계는 각 기능별로 서로간의 독립성을 유지하는 설계를 창출하는 방법이기 때문에, 이를 소프트웨어 분야에 적용할 수 있다. 이에 최근에는 소프트웨어 개발 분야에 공리적설계방법의 개념이 성공적으로 도입되고 있다. 이 절에서는 공리적설계방법을 적용하여 기존의 소프트웨어 공학이 가지고 있는 한계를 극복할 수 있는 방법에 대하여 설명한다.

2.5.2 전통적인 프로그래밍 언어들과 설계공리

소프트웨어 공학에서는 복잡성을 제어하는 가장 좋은 방법으로 분할에 의한 개념이 사용되고 있다. 즉, 프로그램을 작고 이해하기 쉬운 모듈로 분할하는 것이다. 그러나 모듈이 독립적이고 상호 작용이 제한되어 있지 않으면, 복잡성은 제어될 수 없다. 모듈화는 크고 복잡한 시스템을 설계할 때 채택하는 방법인데 시스템을 기능 단위로 분해시켜 주고 수정, 유지보수를 용이하게 한다.

공리적설계 관점에서 소프트웨어의 설계를 표현하기 위해 소비자요구, 기능요구, 설계파라미터, 생산변수들의 특성을 다음과 같이 재정의할 수 있다.

CAs : 소프트웨어가 만족해야 하는 고객의 요구 사항이나 속성
FRs : 공학적인 용어로 표현된 소프트웨어에 대한 요구사항이나 요구사항을 실현하기 위하여 필요한 기능
DPs : (1) 알고리즘을 개발할 경우에는 입력 데이터
(2) 소프트웨어가 적재되는 하드웨어로부터의 신호
(3) 프로그램 코드
PVs : 서브루틴, 기계어, 컴파일러

위에서 정의한 내용을 바탕으로 공리적설계의 관점에서 소프트웨어를 개발하는 과정을 각 단계별로 설명한다. 도서관리용 소프트웨어 시스템 개발의 간단한 예제에 적용하여 이해를 돕고자 한다(Kim *et al.*, 1991 ; Suh, 2000).

단계 1. 소프트웨어의 기능요구들을 정의(Definition of FRs for the software system)

고객의 요구에 따라 기능적 영역에서 최상위수준의 기능요구들을 정의한다. 하나의 기능요구는 소프트웨어가 추구해야 할 기능을 표현한다. 설계 데이터를 입력받고, 프로세스를 실행하거나 어떠한 값을 계산하는 등의 기능을 정의하는 것으로 일반적으로 동사로 시작한다.

도서관리용 소프트웨어 시스템에서의 기능요구는 아래와 같다.

FR_1 : 새로이 반입된 책들에 대한 식별번호와 핵심어를 생성하여라.
FR_2 : 주제별 검색에 대한 도서 리스트를 제공하여라.

단계 2. 기능요구들의 독립성을 유지하는 하에서의 영역간 사상 과정(Mapping between the domains to maintain the independence of FRs)

기능요구를 실행하기 위한 설계파라미터를 물리적 영역에서 설정한다. 설계파라미터는 기능요구를 만족시키기 위해서 '어떻게' 할 것인가를 말한다. 소프트웨어 설계에서 설계파라미터는 입력 데이터와 프로그램을 실행하여 전달하는 결과 데이터에 해당한다. 그러므

로 일반적으로 설계파라미터는 데이터 명칭의 명사로 정의한다. 기능요구들의 독립성은 기능요구와 설계파라미터의 관계에 의해 도출한 설계방정식의 형태로 점검할 수 있다.

단계 1에서 정의한 각 기능요구에 대한 설계파라미터를 아래와 같이 선정할 수 있다.

DP_1 : 각 책들의 목차를 바탕으로 한 분류 시스템

DP_2 : 주제별 핵심어들을 바탕으로 한 검색 시스템

이 과정은 식 (2·79)와 같이 비연성화설계로 순서에 따른다면 독립공리를 만족한다.

$$\begin{bmatrix} FR_1 \\ FR_2 \end{bmatrix} = \begin{bmatrix} A_{11} & 0 \\ A_{21} & A_{22} \end{bmatrix} \begin{bmatrix} DP_1 \\ DP_2 \end{bmatrix} \tag{2·79}$$

소프트웨어 설계에서 A_{ij}는 연산이나 하나의 호출 함수일 수 있다. 이때 기능요구는 식 (2·80)과 같이 모듈에 의해 만족시킬 수 있다.

$$\begin{aligned} FR_1 &= M_1 DP_1 \\ FR_2 &= M_2 DP_2 \end{aligned} \tag{2·80}$$

여기서 M_1과 M_2는 모듈로 다음과 같이 정의된다.

$$\begin{aligned} M_1 &= A_{11} \\ M_2 &= A_{21} \frac{DP_1}{DP_2} + A_{22} \end{aligned} \tag{2·81}$$

모듈은 알고리즘으로 볼 수 있는데 작은 단위로는 논리연산이 있을 수 있고, 크게는 하나의 독립적인 시스템을 구성하는 함수가 될 수 있다.

단계 3. 각 기능요구와 설계파라미터의 분해 과정(Decomposition of FRs and DPs)

앞 절에서 최상위수준에서의 기능요구들과 설계파라미터들은 최하위수준까지 분해 과정을 진행한다고 설명하였다. 분해 시 현 수준에서의 각 설계파라미터는 하위수준의 기능요구들로 분해하기 위한 기준이 된다. 현 수준에서 기능요구 FR_i를 실행하기 위한 설계파라미터 DP_i를 어떠한 것으로 설정하느냐에 따라서 하위수준의 기능요구 FR_{ij}를 다르게 정의한다. 그러므로 기능요구와 설계파라미터를 하위수준으로 분해하는 방식은 지그재그 과정으로 진행된다.

앞 단계에서 예를 들었던 도서관리용 소프트웨어 시스템의 FR_1은 FR_{11}, FR_{12}로 분해할 수 있다.

FR_{11} : 새로운 도서에 식별번호를 부여하여라.

FR_{12} : 새로운 도서의 주제별 핵심어를 생성하여라.

DP_{11} : 도서의 표제 정보

DP_{12} : 도서의 목차

FR_2도 마찬가지로 분해할 수 있고, 최하위수준까지 분해 과정을 진행한다. 예제의 시스템에 대해서 최하위수준까지 분해한 결과는 그림 **2 · 29**와 같다.

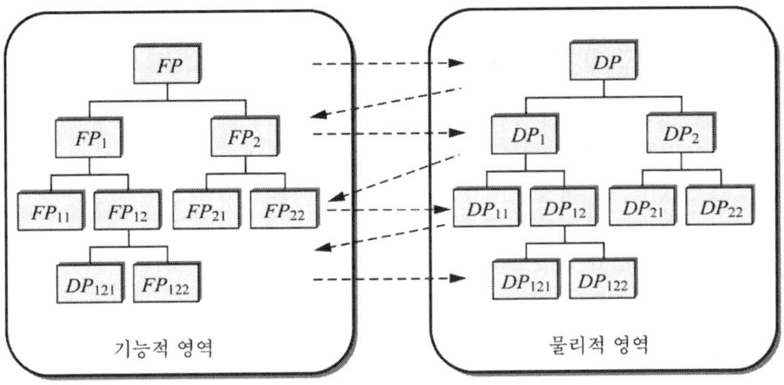

그림 2 · 29 도서관리용 소프트웨어 시스템의 계층구소

단계 4. 모듈 정의 (Definition of Modules)

분해 과정 후에는 모든 기능요구와 설계파라미터에 따르는 모듈을 구성한다. 각 모듈은 독립적으로 프로그램화할 수 있다. 각각의 모듈들을 조합하는 방법은 기능요구의 가지들과 그 가지의 수직 통합을 도표 형태로 표시하는 방법을 이용하며, 구체적인 형태로는 결합(junction)과 모듈(module)로 구성한다. 그림 **2 · 30**에서는 단위 결합의 형태에 따른 표현을 보이고 있다. 세 가지의 결합형태가 있을 수 있으며, 각각의 형태는 다음과 같다.

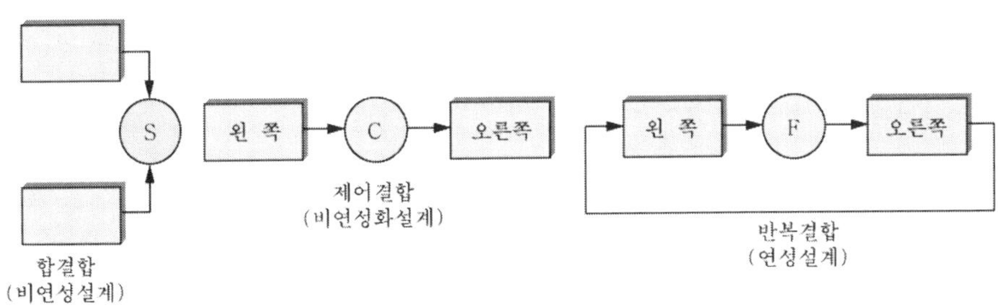

그림 2 · 30 단위 결합의 형태

합결합(summation junction : Ⓢ)은 비연성설계의 경우로 상위수준의 기능요구가 하위 수준 모듈의 모든 결과를 결합함으로써 만족되는 형태이다.

제어결합(control junction : Ⓒ)은 비연성화설계의 경우로 그림 **2 · 30**에서 오른쪽 모듈의 행위를 제어하기 위해 왼쪽 모듈의 결과를 사용하는 형태이다.

반복결합(feedback junction : Ⓕ)은 연성설계의 경우로 그림 **2 · 30**에서 오른쪽 모듈의 결과를 왼쪽 모듈로 되돌려 보낸다. 그러므로 정보의 진행 과정이나 프로그램의 이용에 있어서 수많은 반복을 필요로 한다. 반복결합이 많을 때 그 프로그램은 곧 감당하기 어렵게 된다.

이들 결합 특성을 이용하여 기능요구와 설계파라미터의 계층적 트리구조(tree structure)와 설계행렬의 형태로 나타난 해석의 결과는 하나의 도표로 나타낼 수 있다. 이를 모듈결합구조도(module junction structure diagram)라고 하며 그림 **2 · 29**의 예제에 대해서 그림 **2 · 31**과 같이 표현할 수 있다. 또한 결합 특성을 사용함으로써 모듈결합구조도는 모듈들 간의 정보흐름을 보여주는 네트워크 형태의 그림으로 바뀌어지기 쉽다. 그림 **2 · 32**는 그림 **2 · 31**에서 유추한 흐름도이다.

이 절에서는 소프트웨어를 개발하기 위하여 공리적설계의 설계 과정에 따라 소프트웨어의 모듈을 구성하고, 모듈을 코딩하는 개발 과정을 설명하였다. 전통적인 프로그래밍 언어로 소프트웨어를 개발할 때 공리적설계의 독립공리를 이용하면 모듈의 기능적인 독립성을 확보할 수 있어 소프트웨어 개발 및 유지보수가 쉽다.

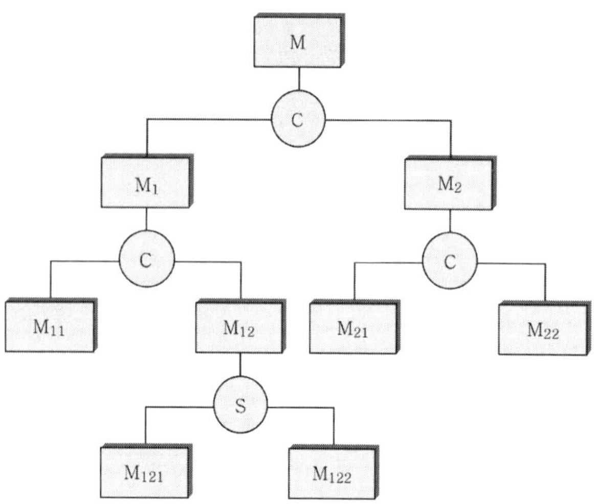

그림 2 · 31 그림 2 · 29의 모듈 결합도

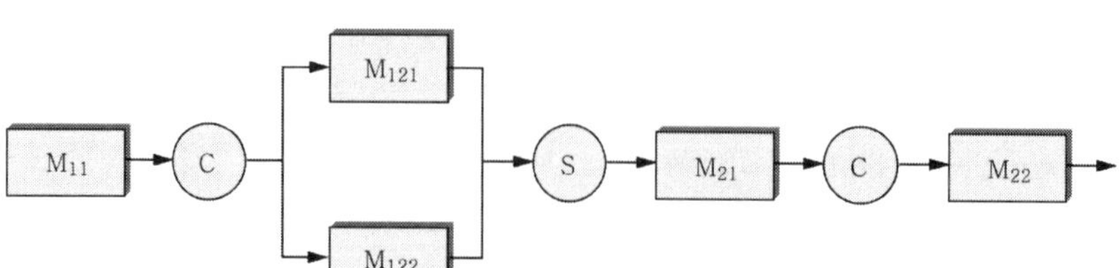

그림 2 · 32 그림 2 · 29의 흐름도

2.5.3 객체 지향적 프로그래밍과 설계공리

소프트웨어 공학에서는 1980년대 후반부터 객체지향 패러다임이 많은 호응을 얻었다. C, Pascal, Fortran과 같은 절차지향(process oriented) 프로그래밍과는 달리 최근 소프트웨어 프로젝트의 주류를 이루고 있는 사용자 중심의 대화형 프로그램을 개발하는 데에는 객체지향 프로그래밍 기술이 우수한 방법을 제공한다고 알려져 있다. 객체지향 기술은 새로운 프로그램의 개발 시 이전 프로그램을 재사용할 수 있는 방법을 제공한다. 즉, 공통으로 사용할 수 있는 부분을 라이브러리로 만들어 필요할 때 적절히 특수화(specialization)하고 알맞게 변경하여 사용할 수 있다. 객체지향 기술의 또 다른 특징은 프로그램을 뚜렷하게 구별되는 객체(object)로 분할할 수 있다는 것이다. 구별한 객체들은 잘 정의된 인터페이스를 이용하여 상호 협조할 수 있다.

이러한 장점으로 최근에 객체지향 프로그래밍(object-oriented programming, OOP)이 소프트웨어 개발에 많이 도입되고 있다. 이에 설계 과정의 기능적 독립성을 객체지향적 프로그래밍에 도입한 V-모델이 제안되었다(Do and Suh, 2000). V-모델에서 설계자는 소프트웨어의 기능을 정의하고 지그재그 과정을 통하여 분해하여 물리적 영역에서의 독립적 모듈을 정한다. 다음에 각 모듈을 객체지향적 프로그래밍의 클래스로 정의하고 프로그램으로 코딩(coding)한다. 이는 크게 하향식 접근방법에 의한 설계행렬(full design matrix)의 구성과 상향식 접근방법에 의한 소프트웨어 완성의 두 단계로 구분한다. 위에서 설명한 과정을 세분화하여, 그림 2·33과 같은 몇 가지 단계로 구분할 수 있다. 그림에서 단계 4까지의 하향식 과정은 앞 절에서 설명한 소프트웨어 개발 과정과 같다. 이후에 전체 설계행렬을 이용하여 객체를 정의하고 코딩을 하는 단계 5부터 설명한다.

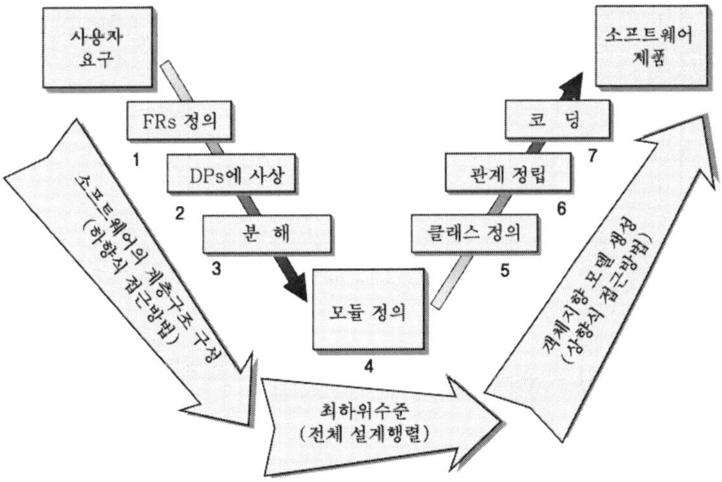

그림 2·33 객체 지향적 소프트웨어 시스템을 위한 공리적설계 과정(V-모델)

단계 5. 객체, 속성, 연산 식별(Identification of objects, attributes and operations)

분해 과정 후에는 모든 기능요구와 설계파라미터의 관계를 보이는 전체 설계행렬(full design matrix)을 구성한다. 그림 **2 · 34**에 전체 설계행렬의 예가 있다. 그림에서 설계행렬의 가로 행은 기능요구, 세로 열은 설계파라미터를 나타내고 'X'는 설계파라미터의 정보로 기능요구를 수행하는 논리적인 관계를 표현하는 알고리즘을 의미한다. 논리관계는 '+'나 '*'와 같은 연산자뿐만 아니라 'if 문'이나 'for 문'과 같은 제어 및 순환 관계 등을 포괄적으로 포함한다. 전체 설계행렬을 객체 지향적 관점에서 바라본다. 그림 **2 · 34**에서 설계행렬의 가로행을 따라 굵은 선으로 표현한 사각형은 하나의 객체를 생성한다. 객체는 세로 열의 설계파라미터에 해당하는 속성(attributes)과 굵은 사각형 내에 존재하는 연산 리스트의 메소드(methods)로 구성된다. 객체의 메소드는 모듈에 해당한다. 그러므로 객체는 주어진 속성을 가지고 메소드를 실행한다. 메소드를 실행함으로써 기능요구를 만족시킬 수 있다.

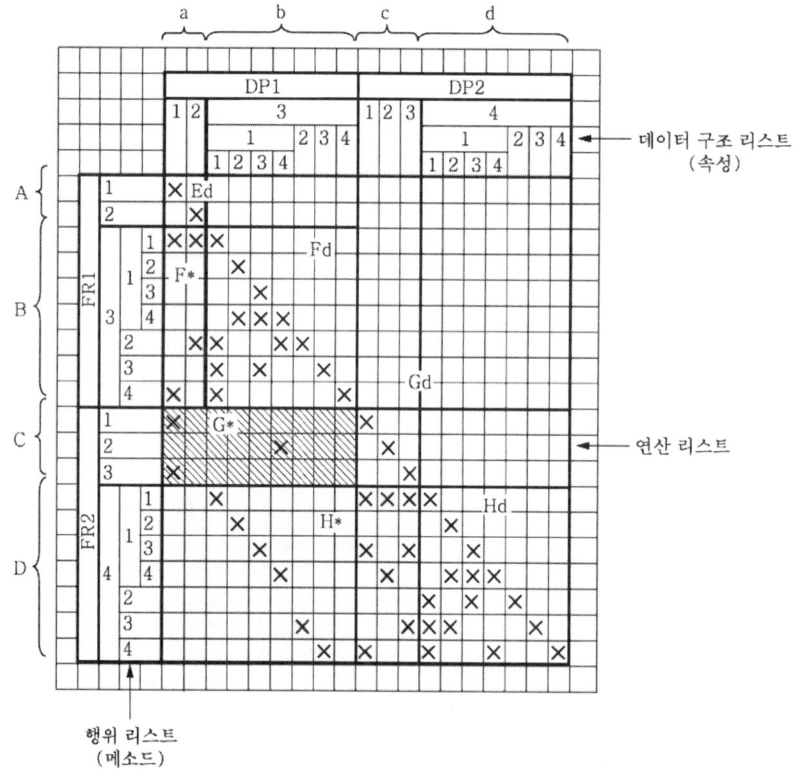

그림 2 · 34 객체 지향적 프로그래밍의 용어를 이용한 전체 설계행렬의 표현

단계 6. 객체들의 관계 정립(Establishment of interfaces between objects)

객체는 클래스에 의하여 표현할 수 있는데, 클래스는 객체의 형식을 정의하는 템플릿(template)이다. 이와 같이 클래스는 일정한 형식을 가진 객체들을 대표하는 것으로 데이터 구조인 속성과 공통적인 행위인 메소드들을 공유한다. 이 단계는 정의된 클래스들 사이의 관계를 설정하는 단계이다. 일반적으로, 그 관계는 일반화(generalization), 집단화(aggregation), 결합화(association)로 구분한다. 그림 **2·35**는 그림 **2·34**의 설계행렬에 따른 클래스 사이의 관계를 클래스 다이어그램으로 보여준다. 전체 설계행렬에 의해 설계에 필요한 데이터 및 기능을 알 수 있다. 또한 앞에서 설명한 설계 흐름도를 이용하면 설계 과정을 알 수 있다. 이들을 이용하여 소프트웨어 개발을 쉽게 진행할 수 있다.

단계 7. 시스템 구성에 따른 코딩작업(Coding with system architecture)

코딩(coding)은 클래스들과 그들 사이의 관계를 근거로 프로그램을 작성하는 것이다. 작업 중에 모듈 사이의 관계와 설계 과정은 모듈 결합도와 설계행렬의 흐름도를 통하여 도움을 얻는다.

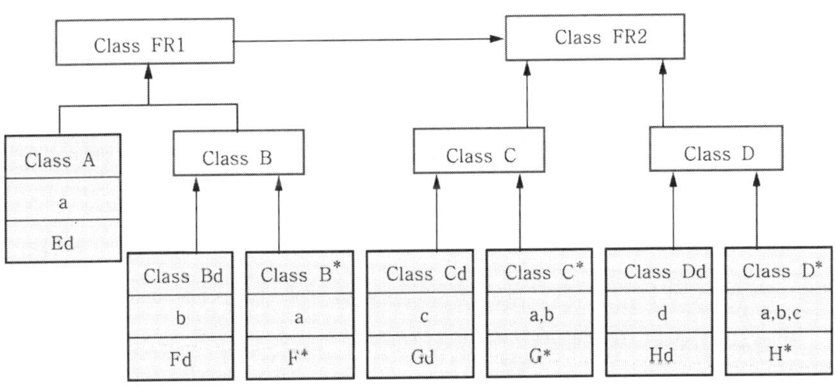

그림 2·35 그림 2·34의 전체 설계행렬에 대한 클래스 다이어그램

2.6 고 찰

앞에서 설명한 바와 같이 독립공리와 정보공리는 서로 독립적인 관계이다. 따라서 두 공리는 따로 적용해야 한다. 대체적으로 독립공리를 만족시키는 것이 우선적인 과제이다. 많은 경우 독립공리의 적용만으로 설계를 종료하기도 한다. 독립공리와 정보공리를 모두 적용할 때는 그림 **2·20**에 보이는 흐름도를 권장한다.

독립공리를 적용할 때에는 이상적인 설계를 염두에 두는 것이 좋다. 이상적인 설계는 기능요구의 수와 설계파라미터의 수가 동일해야 한다. 설계행렬은 정방형의 대각행렬이나 삼각행렬이어야 한다. 양쪽의 숫자가 다른 경우는 모두 연성설계가 된다. 설계파라미터의 숫자가 모자랄 때는 새로운 설계파라미터를 추가해야 한다. 설계파라미터의 숫자가 더 많은 과잉설계의 경우에는 숫자를 줄이거나, 설계행렬을 관찰하면서 특정 파라미터를 고정시킬 수 있다.

예를 들어 다음과 같은 과잉설계를 보자.

$$\begin{bmatrix} FR_1 \\ FR_2 \end{bmatrix} = \begin{bmatrix} X & X & 0 \\ 0 & X & X \end{bmatrix} \begin{bmatrix} DP_1 \\ DP_2 \\ DP_3 \end{bmatrix} \qquad (2 \cdot 82)$$

우선 DP_2를 고정시킨다. 그러면 나머지의 파라미터로 만드는 설계행렬은 비연성설계가 된다. 즉, 과잉의 파라미터를 고정시키되 나머지의 파라미터로 비연성설계나 비연성화설계가 되도록 한다.

정보공리는 독립공리를 만족한 설계를 정량적으로 평가하는 데 사용한다. 특히 독립공리를 만족하는 설계안이 복수일 때 어느 것이 더 좋은 설계인가를 판별하는 데 유용하다.

이상으로 공리적설계를 설명하였다. 기본적으로 공리적설계는 새로운 설계를 창출하거나 우수한 설계를 판별하는 데 유용하게 사용할 수 있다. 특히 기존에 없던 제품을 새로이 만들 경우 개념설계를 행하는 데 대단히 효과적이다. 공리적설계는 다른 설계방법에 비해 비교적 최근에 개발되었으나, 이미 많은 설계 예를 통해 그 유용성이 입증되었다(Suh, 1990 · 2000). 공리적설계를 이용하는 설계자들로부터 공통적으로 관찰되는 현상이 있다. 공리를 처음으로 접하였을 때는 공리에 대한 공감대가 쉽게 형성되고 즉시 설계에 적용할 수 있다고 생각한다. 그리고는 현재 다루고 있는 제품에 공리를 적용하려 한다. 그러나 대부분의 경우 기존 제품을 공리적설계로 분석하는 과정에서 커다란 어려움을 느끼게 된다. 그 이유는 제품을 기존의 설계관점으로만 생각하고 공리적설계의 관점으로 분석하지 않기 때문이다. 이 과정에서 대부분의 설계자들이 어려움을 느끼게 되고 그만두는 경우가 발생한다. 그러나 이 과정을 극복하면 공리적설계의 유용성을 인지하게 된다. 어려움을 극복하는 데 권장할 사항은 기능요구를 정의할 때 기존 제품을 의식하지 않아야 한다는 것이다. 그 대신 순수하게 무엇을 하고자 하는가의 기능요구만을 중립적 상태(solution neutral environment)에서 깊이 생각해야 한다. 특히 많은 설계자들은 설계를 물리적 영역에서 생각하는 경향이 있는데 공리적설계는 기능적 영역에서 생각하기를 권장한다.

최근의 연구에서 공리적설계를 상세설계에 이용하고 있다. 구조최적설계에서 구조물의 치수를 정하는 데 응용하는 예가 뒤에서 보일 것이다.

참 고 문 헌

강병수 · 송기남 · 최성규 · 박경진 · 윤경호, 「공리적설계를 이용한 원자로 핵연료봉 지지격자체의 설계」, 『대한기계학회논문집』 A 26(8), pp. 1623~1630, 2002.

강영주 · 황윤동 · 차성운 · 정대진 · 문용락, 「공리적설계에서 FRs의 상호 독립성을 정량적으로 측정하기 위한 도구의 개발」, 『한국정밀공학회지』 18(3), pp. 122~130, 2001.

도성희, 「설계공리를 이용한 TV 유리벌브의 제조를 고려한 설계방법 개발」, 박사학위논문, 한양대학교, 1997.

도성희 · 박경진, 「설계 공리를 이용한 유리벌브 제품설계 자동화 시스템 구축」, 『대한기계학회논문집』 A 20(4), pp. 1333~1346, 1996.

박경진 · 도성희 · 서남표, 「공리적설계 방법을 이용한 소프트웨어 시스템의 설계 및 확장 – TV Glass Bulb 설계 소프트웨어를 중심으로–」, 『대한기계학회논문집』 A 23(9), pp. 1536~1549, 1999.

박경진 · 도성희 · 이정욱 · 김창훈 · 김재선, 「제품설계 자동화 시스템 구축」, 프로젝트 최종 보고서, 삼성코닝(주), 1995.

신광섭 · 박경진, 「레이저 장치의 보조빔 정렬장치」, 대한민국 특허 등록 제10-0453984호, 2004.

신광섭 · 이정욱 · 박경진, 「레이저 마커용 빔 정렬장치의 공리적설계」, 『대한기계학회논문집』 A 26(9), pp. 1727~1735, 2002.

신문균 · 홍성우 · 박경진, 「안전도를 고려한 전동 틸트/텔레스코픽 조향주의 해석 및 설계」, 『대한기계학회논문집』 A 24(6), pp. 1479~1490, 2000.

이광원, 「독립 설계 공리를 이용한 구조물의 최적설계」, 박사학위논문, 한양대학교, 1998.

이광원 · 박경진, 「독립 공리를 이용한 구조 최적화방법론 개발」, 『대한기계학회논문집』 A 24(10), pp. 2438~2450, 2000.

이정욱 · 하대율 · 이상우 · 임재문 · 박경진, 「공리적설계를 이용한 모니터용 EPS 완충 포장 설계 시스템 개발」, 『대한기계학회논문집』 A 27(10), pp. 1644~1652, 2003.

차성운 · 박경진 역, 『공리적설계 [I]』, 동명사, 2002.

황광현 · 박경진, 「강건성을 고려한 공리적설계의 새로운 정보 지수」, 『대한기계학회논문집』 A 26 (10), pp. 2073~2081, 2002.

황광현 · 이권희 · 박경진 · 이병렬 · 조용철 · 이석한, 「공리적설계를 이용한 비대칭 내부 짐벌을 가진 진동형 자이로스코프의 강건설계」, 『대한기계학회논문집』 A 26(5), pp. 914~923, 2002.

Albano LD & Suh NP, "Axiomatic Approach to Structural Design," *Research in Engineering Design* 1(4), pp. 171~183, 1992.

Albano LD & Suh NP, The Information Axiom and its Implication, DE-Vol 66, *Intelligent Concurrent Design: Fundamentals, Methodology, Modeling and Practice, ASME*, 1993.

ANSYS Inc., *ANSYS User's Manual Version 8.1*, Canonsburg, PA, 2004.

Do SH & Park GJ, "Application of Design Axioms for Glass Bulb Design and Software Development for Design Automation," *Journal of Mechanical Design, Transaction of the ASME* 123(3), pp. 322~329, 2001.

Do SH & Suh NP, "Axiomatic Design of Software Systems," *CIRP Annals* 49(1), pp. 95 ~ 100, 2000.

Electronic Data Systems Co., *User Function Programming Manual*, Livonia, MI, 1993.

El-Haik B & Yang K, "The Components of Complexity in Engineering Design," *IIE Transactions* 31, pp. 925~934, 1999.

Frey DD & Jahangir, E & Engelhardt, F., "Computing the Information Content of Decoupled Designs," *Research in Engineering Design* 12, pp. 90~102, 2000.

Gebala DA & Suh NP, "An Application of Axiomatic Design," *Research in Engineering Design* 3, pp. 149~162, 1992.

Heller D & Ferguson PM, *Motif Programming Manual*, O'Reilly & Associaties, Setastopol, CA, 1994.

Hewlett-Packard Co. *Starbase Graphics Techniques*, TX, 1991.

Hwang KH & Lee KW & Park GJ, "Robust Optimization of an Automobile Rearview Mirror for Vibration Reduction," *Structural and Multidisciplinary Optimization* 21(4), pp. 300~308, 2001.

Kim SJ & Suh NP & Kim SK, "Design of Software Systems Based on Axiomatic Design," *Annals of CIRP*, 40(1), pp. 165~170, 1991.

Lee J & Cho K & Lee K, "A New Control System of a Household Refrigerator-Freezer," International Refrigeration Conference, Purdue University, IN, 1994.

Livermore Software Technology Co. *LS/INGRID Manual Ver.* 3.5, Livermore, CA, 1998.

Livermore Software Technology Co. *LS/DYNA Theory Manual*, Livermore, CA, 1998.

Livermore Software Technology Co. *LS/DYNA User's Manual Ver.* 950, Livermore, CA, 1999.

Mood AM & Graybill FA & Boes DC, *Introduction to the Theory of Statistics*, 3rd ed., McGraw-Hill, New York, 1963.

NSF, *Axiomatic Design Workshop for Professors*, MIT, Cambridge, MA, 1998.

Oracle Co, *ORACLE 3GL Programmers Guide*, Redwood Shore, CA, 1990.

Phadke MS, *Quality Engineering Using Robust Design*, Prentice Hall, Eaglewood Cliffs, NJ, 1989.

Pressman RE, *Software Engineering*, McGraw Hill, New York, 1997.

Rinderle JR & Suh NP, "Measures of Functional Coupling in Design," Transactions of ASME, *Journal of Engineering for Industry* 104, pp. 383~388, 1982.

Shin MK & Hong SW & Park GJ, "Axiomatic Design of the Motor-Driven Tilt/Telescopic Steering System for Safety and Vibration," *Proceeding of the Institution of Mechanical Engineers, Part D – Journal of Automobile Engineering* 215(2), pp. 179~187, 2001.

Suh NP, *The Principles of Design*, Oxford University Press, New York, 1990.

Suh NP, "Axiomatic Design of Mechanical System, Special 50th Anniversary, Combined Issue of the Journal of Mechanical Design and the Journal of Vibration and Acoustics," *ASME* 17, pp. 1~10, 1995.

Suh NP, "Designing-in of Quality through Axiomatic Design," *IEEE Transactions on Reliability* 449(2), pp. 256~264, 1995.

Suh NP, "Design and Operation of Large Systems," *Journal of Manufacturing Systems* 14(3), pp. 203~213, 1995.

Suh NP, "Axiomatic Design Theory for Systems," *Research in Engineering Design* 10, pp. 189~209, 1998.

Suh NP, "A Theory of Complexity, Periodicity and the Design Axioms," *Research in Engineering Design* 11, pp. 116~131, 1999.

Suh NP, *Axiomatic Design: Advances and Applications*, Oxford University Press, New York, 2000.

Suh NP & Bell AC & Gossard DC, "On an Axiomatic Approach to Manufacturing and Manufacturing Systems," *Journal of Engineering for Industry* 100(2), pp. 127~130, 1978.

Suh NP & Sekimoto, S., "Design of Thinking Design Machine," *Annals of the CIRP* 39(1), pp. 145~148, 1990.

Suh NP & Wilson DR & Tice WW & Yasuhara M & Bell AC, "Application of Axiomatic Design Techniques to Manufacturing, Winter Annual Meeting," ASME 79-WA/Prod-25, New York, 1979.

Taguchi G, *Systems of Experimental Design*, Kraus International Publications, New York, 1987.

Yi JW & Park GJ, "Development of a Design System for EPS Crushing Package of a Monitor Using Axiomatic Design," *Advances in Engineering Software* 36, pp. 273~284, 2005.

연 습 문 제

1. CD 플레이어 설계를 공리적설계의 독립공리의 관점에서 분석하여라.
2. 휴대전화의 설계를 공리적설계의 독립공리의 관점에서 분석하여라.
3. 주위에서 물리적통합의 아이디어를 사용한 제품을 찾아 공리적설계의 독립공리의 관점에서 분석하여라.
4. 자동차의 연료탱크를 설계하려고 하는데 다음의 기능요구가 설정되었다.

   FR_1 : 탱크에 연료를 주입할 수 있게 하여라.
   FR_2 : 탱크가 가득 차면 연료 주입이 멈추도록 하여라.
   FR_3 : 탱크 내의 연료의 수위가 주입관보다 높으면 연료의 기압에 의해 역류할 수 있는데 이를 방지하여라.
   FR_4 : 탱크 내의 연료에 의한 기압을 조절하여라.

   위의 네 가지 기능요구를 만족하는 연료탱크를 설계하여라. 단 탱크의 가격이 현재 가격보다 높지 않게 하여라.

5. FR-DP의 관계가 다음과 같은 설계행렬이 있다.

$$\begin{bmatrix} f_1 \\ f_2 \\ f_3 \end{bmatrix} = \begin{bmatrix} X & 0 & 0 \\ X & X & 0 \\ X & X & X \end{bmatrix} \begin{bmatrix} x_1 \\ x_2 \\ x_3 \end{bmatrix}$$

   그들의 자세한 관계는

$$f_1 = x_1^2$$
$$f_2 = x_1^2 + x_1 x_2 + x_2^2$$
$$f_3 = x_1^2 + x_2^2 + x_2 x_3 + x_3^2$$

   이다.

   (1) $x = [\,1.0 \;\; 1.0 \;\; 1.0\,]^T$일 때의 근사화된 설계행렬을 구하여라.
   (2) 위의 설계가 특정한 설계창에서 비연성설계가 되기 위한 조건을 구하여라.

6. 다음의 설계행렬에 대한 리앵귤러리티와 셈앵귤러리티를 구하고 설계행렬의 특성을 논하여라.

   (a) $\begin{bmatrix} 3 & 1 \\ 1 & 2 \end{bmatrix}$ (b) $\begin{bmatrix} 8 & 0.2 \\ 0.3 & 4 \end{bmatrix}$ (c) $\begin{bmatrix} 9.5 & 2.4 \\ 3.2 & 4.8 \end{bmatrix}$

   (d) $\begin{bmatrix} 5 & 0 \\ 1 & 2 \end{bmatrix}$ (e) $\begin{bmatrix} 5 & 0 \\ 4 & 2 \end{bmatrix}$ (f) $\begin{bmatrix} 3 & 0 \\ 0 & 2 \end{bmatrix}$

7. 다음의 두 가지 설계안이 있다.

$$\text{설계안 1} \begin{bmatrix} FR_1 \\ FR_2 \end{bmatrix} = \begin{bmatrix} 3 & 1 \\ 2 & 4 \end{bmatrix} \begin{bmatrix} DP_1 \\ DP_2 \end{bmatrix}$$

$$\text{설계안 2} \begin{bmatrix} FR_1 \\ FR_2 \end{bmatrix} = \begin{bmatrix} 3 & 0 \\ 8 & 5 \end{bmatrix} \begin{bmatrix} DP_1 \\ DP_2 \end{bmatrix}$$

(1) 각 설계에 대한 FR-DP의 그래프를 그리고 그래프 상에서 설계가 진행되는 순서를 설명하여라.

(2) 각 설계안에 대한 리앵귤러리티(R)와 셈앵귤러리티(S)를 구하고 결과를 비교하여 논하여라.

8. FR-DP의 관계가 다음과 같은 설계행렬이 있다.

$$\begin{bmatrix} FR_1 \\ FR_2 \\ FR_3 \end{bmatrix} = \begin{bmatrix} 5 & 0.1 & 0.2 \\ 0.3 & 7 & 0.4 \\ 0.5 & 0.7 & 4 \end{bmatrix} \begin{bmatrix} DP_1 \\ DP_2 \\ DP_3 \end{bmatrix}$$

설계파라미터가 $DP_1 = DP_2 = DP_3 = 1$의 값을 가질 때 제조 시 설계 공차가 ΔDP_1, ΔDP_2, ΔDP_3이다.

(1) 식 (2·21)에서 $(\Delta FR_i)_{\text{allowable}}$이 0.5일 때 위의 설계가 비연성설계가 되기 위한 조건을 ΔDP_1, ΔDP_2, ΔDP_3로 이루어진 연립부등식으로 나타내어라.

(2) 설계자가 지정한 기능요구의 범위가 $4.8 < FR_1 < 5.8$, $7.3 < FR_2 < 8.0$, $5.1 < FR_3 < 6.3$이라 할 때 위의 설계를 비연성설계로 만드는 조건을 문제 (1)과 같은 방법으로 나타내어라.

9. 다음의 세 가지 설계안이 있다.

$$\text{설계안 1} \begin{bmatrix} FR_1 \\ FR_2 \end{bmatrix} = \begin{bmatrix} 3 & 0 \\ 0 & 4 \end{bmatrix} \begin{bmatrix} DP_1 \\ DP_2 \end{bmatrix}$$

$$\text{설계안 2} \begin{bmatrix} FR_1 \\ FR_2 \end{bmatrix} = \begin{bmatrix} 3 & 0 \\ 2 & 4 \end{bmatrix} \begin{bmatrix} DP_1 \\ DP_2 \end{bmatrix}$$

$$\text{설계안 3} \begin{bmatrix} FR_1 \\ FR_2 \end{bmatrix} = \begin{bmatrix} 3 & 1 \\ 2 & 4 \end{bmatrix} \begin{bmatrix} DP_1 \\ DP_2 \end{bmatrix}$$

(1) 위의 설계안 중 독립공리를 만족하는 것은?

(2) 독립공리를 만족하는 설계 중 리앵귤러리티와 셈앵귤러리티의 계산을 통해 독립성이 우수한 것을 판별하여라.

(3) 독립공리를 만족하는 설계에 대해서 $\Delta DP_i = 0.1$, $\Delta FR_i = 0.5$, $i = 1, 2$이다. 기능요구가 만족되는 설계점에서 정보량을 구하여라.

(a) DP영역에서 DP가 균등분포를 가질 때 도식적 방법으로 각각 구하고 결과를 비교하여라.

(b) FR영역에서 (a)와 동일한 조건하에서 구하여라.

(c) 설계파라미터의 분포가 다음과 같을 때

$$DP_1\text{의 분포함수 } p_{\delta DP_1} = -\frac{3}{4(\Delta DP_1)^3}(\delta DP_1)^2 + \frac{3}{4(\Delta DP_1)}$$

$$DP_2\text{의 분포함수 } p_{\delta DP_2} = \frac{1}{(\Delta DP_2)^2}\delta DP_2 + \frac{1}{\Delta DP_2},\ \delta DP_2 < 0\text{일 때}$$

$$p_{\delta DP_2} = \frac{-1}{(\Delta DP_2)^2}\delta DP_2 + \frac{1}{\Delta DP_2},\ \delta DP_2 \geq 0\text{일 때}$$

적분법에 의해 각각에 대해 구하고 비교하여라.

10. 다음의 비연성화설계가 있다. 설계행렬의 한 요소가 다음과 같이 미지수 x로 표기되어 있다. x가 0.5, 1, 2의 값을 가질 때

$$\begin{bmatrix} 0 < FR_1 < 2 \\ 0 < FR_2 < 2 \end{bmatrix} = \begin{bmatrix} 1 & 0 \\ 1 & x \end{bmatrix} \begin{bmatrix} 0 < DP_1 < 1 \\ 0 < DP_2 < 1 \end{bmatrix}$$

각각의 x값에 대해

(1) 성공의 확률을 계산하여라.

(2) 성공의 확률이 100%가 되기 위한 설계파라미터의 공차 크기를 계산하여라. 성공의 확률 계산 시 어느 방법이나 사용해도 무방하다. x의 크기에 따른 추세를 논하여라.

11. 집을 사는 문제와 비슷하게 노트북을 사는 문제를 만들어서 풀어 보아라.

12. 두 가지 기능요구를 만족하는 설계를 다음과 같이 발견하였다.

$$\text{설계안 1} \begin{bmatrix} FR_1 \\ FR_2 \end{bmatrix} = \begin{bmatrix} 100 & 0 \\ 0 & 100 \end{bmatrix} \begin{bmatrix} DP_1 \\ DP_2 \end{bmatrix}$$

$$\text{설계안 2} \begin{bmatrix} FR_1 \\ FR_2 \end{bmatrix} = \begin{bmatrix} 1 & 0 \\ 1 & 1 \end{bmatrix} \begin{bmatrix} DP_1 \\ DP_2 \end{bmatrix}$$

여기서 각 FR의 목표값은 10, 20이고 각 FR의 허용범위는 $-9.5 < FR_1 < 10.5$, $-19.5 < FR_2 < 20.5$이며 각 DP의 공차는 ±2.0이다.

(1) 위의 두 설계의 R, S를 구하고 독립성의 관점에서 볼 때 어느 것이 좋은 설계인가?

(2) 각 설계를 만족했을 때 정보량을 도식적으로 구하고 어느 것이 좋은 설계인가를 결정하여라.

(3) 위의 결과가 주는 시사점을 논하여라.

13. 비연성화설계도 설계행렬의 비대각항이 작을 경우에는 비연성설계로 간주할 수 있다. 이를 식 (2·21)과 R값의 관계로 설명하여라.

14. 비연성화설계에서 도식적 방법으로 정보량을 구하는 과정을 연성설계로 확장하여라.

2.A 추론 및 정리

추론 2.A · 1 연성설계의 비연성화

제안한 설계의 기능요구들이 연성되어 있거나 상호 종속적이면 설계해의 부품 혹은 특성을 비연성화시키거나 분리시킨다.

추론 2.A · 2 기능요구 최소화

기능요구와 제한조건의 숫자를 최소화한다.

추론 2.A · 3 물리적 통합

제안한 설계가 기능요구들을 독립적으로 만족할 수만 있다면 가능한 한 설계파라미터들을 하나의 물리적 부품에 통합한다.

추론 2.A · 4 표준화

기능요구와 제한조건을 만족할 수 있으면 표준화되어 있거나 교환 가능한 부품을 사용해야 한다.

추론 2.A · 5 대칭성의 이용

기능요구와 제한조건을 만족할 수 있으면 대칭적인 형상이나 요소를 이용해야 한다.

추론 2.A · 6 최대한의 공차

기능요구를 정의하는 경우 허용할 수 있는 가장 큰 공차를 명시해야 한다.

추론 2.A · 7 가장 적은 정보량을 갖는 비연성설계

기능요구를 만족하는 데 있어서 연성설계보다 적은 정보량이 필요한 비연성설계를 찾는다.

추론 2.A · 8 리앵귤러리티

독립성을 나타내는 설계행렬의 리앵귤러리티 R의 효율적인 값은 1이다.

정리 2.A · 1 설계파라미터 숫자의 부족함으로 인한 연성

설계파라비터의 수가 기능요구의 수보다 적으면 연성설계이거나 기능요구를 만족할 수 없다.

정리 2.A · 2 연성설계의 비연성화

설계파라미터의 수 n과 기능요구의 수 m의 관계가 $m>n$으로 인해 설계가 연성일 때, $n \times n$의 부분행렬이 삼각행렬이 되면 새로운 설계파라미터를 추가하여 기능요구의 수와 설계파라미터의 수를 같게 하여 비연성화할 수 있다.

정리 2.A · 3 과잉설계

기능요구보다 설계파라미터가 많을 때에 설계는 과잉설계이거나 연성설계이다.

정리 2.A · 4 이상적인 설계

이상적인 설계에서 설계파라미터의 수는 기능요구의 수와 동일하다.

정리 2.A · 5 새로운 설계의 필요

새로운 기능요구를 추가하거나 기능요구 중 하나를 교체 혹은 완전히 다른 기능요구 집합을 선정함으로 인해 기능요구의 집합이 바뀌면 원래의 설계파라미터에 의한 설계해는 새로운 기능요구의 집합을 만족시킬 수 없게 된다. 따라서 새로운 설계해를 찾아야 한다.

정리 2.A · 6 비연성설계의 경로 독립성

비연성설계의 정보량은 주어진 기능요구를 만족시키기 위하여 설계파라미터를 변경하는 순서에 대해 독립적이다.

정리 2.A · 7 연성설계와 비연성화설계의 경로 의존성

연성설계와 비연성화설계의 정보량은 설계파라미터를 변경하는 순서와 설계파라미터의 특정 경로에 대하여 종속적이다.

정리 2.A · 8 독립성과 공차

설계자가 지정한 기능요구에 대한 공차가 다음 식보다 크면 설계는 비연성설계이다.

$$\sum_{\substack{j \neq i \\ j=1}}^{n} \left(\frac{\partial FR_i}{\partial DP_j}\right) \Delta DP_j$$

이러한 경우에는 설계행렬의 비대각 원소들을 설계 고려사항으로부터 제외할 수 있다.

정리 2.A · 9 제조성을 고려한 설계

제품의 제조가능을 위해서는, 해당 제품에 대한 설계행렬 A(FR벡터와 DP벡터의 관계)와 제조 공정에 대한 설계행렬 B(DP벡터와 PV벡터의 관계)의 곱이 대각행렬 혹은 삼각행렬이 되어야 한다. 결과적으로 A 혹은 B 중 어느 하나라도 연성설계를 나타내는 경우에 해당 제품은 제조할 수 없다.

정리 2.A · 10 독립성 척도들의 모듈화

설계행렬 A가 주대각항만이 0이 아닌 정방 부분행렬들로 분할될 수 있으면 A의 리앵귤러리티(reangularity)와 셈앵귤러리티(semangularity)로 정의하는 독립성의 척도는 분할된 부분행렬들의 해당 척도들의 곱과 같다.

정리 2.A · 11 불변성

설계행렬 A에 대한 리앵귤러리티와 셈앵귤러리티는 각 FR과 해당 DP의 관계를 유지하는 한 FR과 DP의 순서를 다르게 하더라도 변하지 않는다.

정리 2.A · 12 정보의 합

사건들이 통계적으로 독립이 아닌 경우에 적절한 조건부 확률을 사용하면 사건들의 집합에 대한 정보의 합 역시 정보이다.

정리 2.A · 13 전체 시스템의 정보량

각 설계파라미터가 다른 설계파라미터들에 대하여 확률적으로 독립이면 전체 시스템의 정보량은 각 기능요구에 관련된 정보량의 합이다.

정리 2.A · 14 연성설계 대 비연성설계의 정보량

기능적 영역에서 기능요구의 상태가 변할 때 변화에 필요한 정보는 비연성 과정보다는 연성 과정에서 더 크다.

정리 2.A · 15 설계-제조 관계

제조 시스템이 제품의 기능요구들간의 독립성을 위배하면 제품의 설계를 변경하거나 혹은 새로운 제조 공정을 정의하여 그 독립성을 유지해야 한다.

정리 2.A · 16 정보량의 등가성

설계 작업과 관련된 모든 정보량은 그들의 물리적 근원과 상관없이 동일한 중요성을 가지며 정보량에는 어떠한 가중치도 적용해서는 안 된다.

정리 2.A · 17 완벽한 정보가 없는 경우 설계

비연성화설계의 경우에 부족한 정보가 설계행렬의 비대각 요소들과 관계되면 완벽한 정보가 없는 경우에도 설계는 진행할 수 있다.

정리 2.A · 18 비연성설계의 존재

연성설계보다 작은 정보량을 가지는 비연성설계나 비연성화설계가 항상 존재한다.

정리 2.A · 19 설계의 강건성

독립공리를 만족하는 설계의 정보량을 줄이는 것이 쉽다는 점에서 비연성설계와 비연성화설계는 연성설계보다 강건하다.

정리 2.A · 20 설계영역과 연성

비연성 혹은 비연성화설계의 설계영역이 좁아지면 연성설계가 될 수 있다. 반면 연성설계의 설계영역이 넓어지면 설계는 비연성 혹은 비연성화설계가 될 수 있다.

정리 2.A · 21 설계영역에서 비균등 확률밀도함수를 가질 때의 강건 설계

설계영역에서 FR의 확률밀도함수가 비균등분포를 하면 성공의 확률은 시스템영역이 설계영역 안에 있을 확률이다.

정리 2.A · 22 비연성화설계의 비교 강건성

FRs의 주어진 집합에 대하여 최대 설계 영역이 주어지면 비연성화설계의 DPs에 대한 허용공차는 비연성설계의 허용공차보다 작기 때문에 비연성화설계는 비연성설계만큼 강건할 수 없다.

정리 2.A · 23 비연성화설계의 강건성 감소

설계행렬의 모든 비대각항이 0이 아닌 비연성화설계에서 기능요구의 숫자가 증가하면 허용공차나 강건성이 감소한다.

정리 2.A · 24 최적 계획

로봇 운동계획이나 공장계획에 대한 공정설계는 그것이 최적화되기 전에 연성을 제거해야 한다. 이때 연성은 비연성화기를 이용해 제거할 수 있다. 비연성화기는 하드웨어 혹은 버퍼를 독립시키는 형태이거나 열(queue)의 형태일 수 있다.

정리 2.A · 25 '밀기(push)' 시스템 대 '당기기(pull)' 시스템

동일한 부품이 시스템을 통하여 처리될 때 '밀기' 시스템은 생산성을 최대화하는 비연성화기를 사용하여 설계할 수 있다. 반면 다른 작동들을 요구하는 불규칙한 부품들이 처리될 때는 '당기기' 시스템이 더 효과적이다.

정리 2.A · 26 무한한 시간 동안 조합의 복잡성(combinatorial complexity)을 가지는 시스템의 주기적인 복잡성(periodic complexity)을 가지는 시스템으로의 변환

지속적인 시간에서의 조합의 복잡성을 가지는 시스템의 불확실성은 주기적인 복잡성을 가지는 시스템으로 변환해서 현저히 줄일 수 있다.

2.B 레이저 마커용 빔 정렬장치의 설계

2.B.1 문제의 설명

반도체 집적회로나 산업용 부품의 표면에 로고나 상품정보를 표시하는 기계로 레이저 마커라는 것이 있다(신광섭 외, 2002). 그림 **2.B · 1**은 레이저 마커 중 하나이고 그림 **2.B · 2**는 그 내부를 도식화한 것이다. 이것은 최근에 널리 이용되고 있는 빔스캐닝(beam scanning)식 YAG 레이저 마킹 방법으로, 레이저와 고속으로 제어되는 거울을 이용하여 펜으로 써

나가듯이 표면에 글자를 쓴다. YAG 레이저는 고체 레이저의 하나로 이트륨(yttrium), 알루미늄(aluminium), 가닛(garnet)의 3개 합성물질로 이루어진 결정을 매질로 사용한다. 레이저 마커는 그림 **2.B・1**과 같은 모양으로 크게 YAG 레이저를 발생시키는 빔 발생장치와 스캐닝 헤드부로 구성되어 있다. 스캐닝 헤드는 거울이 부착된 두 쌍의 모터가 장착된 장치로, 거울을 이용하여 레이저빔의 경로를 원하는 위치로 제어한다. 그림 **2.B・2**에서 볼 수 있는 바와 같이 빔 발생 장치에서 생성된 레이저빔은 거울에 반사되어, 그 경로가 스캐닝 헤드부로 입사한다. 이때 마킹 속도를 두 배로 하기 위해서 50%는 반사하고 50%는 투과하는 기능을 가진 거울로 그 경로를 분리한다. 이와 같이 두 군데 위치에서 동시에 마킹할 수 있는 장치를 듀얼형 마커라 하는데 반도체 표면마킹 분야에서 널리 사용하고 있다.

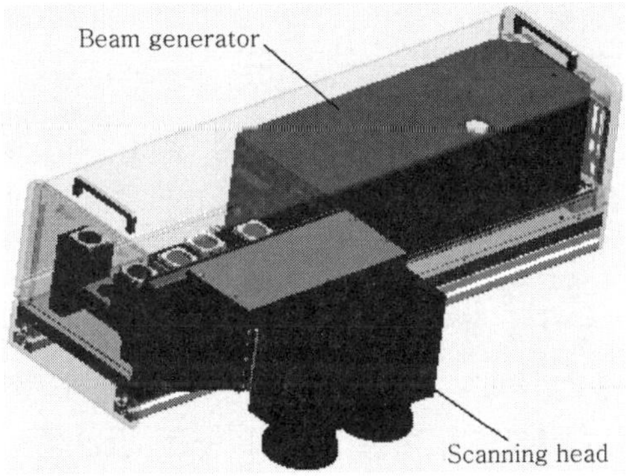

그림 2.B・1 빔스캐닝 타입 레이저 마커

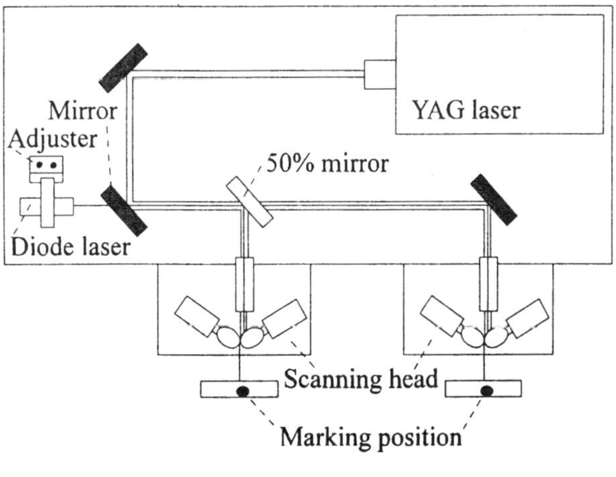

그림 2.B・2 레이저 마커의 내부

실제로 마킹에 사용되는 빔은 YAG 레이저이다, 그러나 여러 가지 새로운 자재를 초기에 위치 설정할 때 고가의 자재를 보호하기 위해서 마킹용 레이저빔을 직접 사용하지 않고 마킹 위치를 확인하고 조정할 수 있어야 한다. 이를 위해 그림 **2.B · 2**의 왼쪽과 같이 저가 저출력의 지시용 다이오드 레이저(diode laser)를 사용하면 마킹 위치를 재설정할 때 자재의 손상 없이 마킹할 경로를 미리 확인하고 보정할 수 있다.

다이오드 레이저를 정렬하는 과정은 다음과 같다. 먼저 YAG 레이저에서 출력된 레이저빔이 스캐닝 헤드부에 입사하도록 정렬시킨다. 그런 뒤 경로를 표시하기 위해 기본적인 마킹 위치에 원점을 표시해 둔다. 다이오드 레이저가 YAG 레이저와 동일한 경로를 따르도록 정렬하기 위해서는 그림 **2.B · 3**과 같이 다이오드 레이저의 표시점과 YAG 레이저로 이미 생성한 마킹 기준점이 같아지도록 다이오드 레이저 빔의 발생 위치를 이동한다. 이때 다이오드 레이저를 마킹용 YAG 레이저의 경로에 정확히 일치시켜야 두 가지 레이저가 같은 위치가 된다. 이렇게 되면 YAG 레이저를 끈 상태에서 다이오드 레이저만으로도 마킹 위치를 안전하게 확인할 수 있다. 마치 사격에서 말하는 영점조정의 기능을 하도록 레이저빔 경로의 두 축을 일치시키는 장치가 빔 정렬장치(beam adjuster)이다.

빔 정렬장치는 마킹 확인용 다이오드 레이저를 이미 정렬된 마킹용 YAG 레이저의 경로에 정확히 일치시키는 역할을 한다. 이때 빔 정렬장치의 수직 · 수평의 위치 및 각도를 조정하여 올바른 마킹 기준점에 정렬해야 한다. 그러나 정렬장치는 가공오차가 아주 미세하여 정렬하기 어려운 경우가 많다. 이러한 문제점은 단순히 정밀한 부품을 사용하거나 많은 조립시간을 투자해도 해결되지 않는 경우가 많다. 그러므로 정렬장치의 위치 및 각도를 정확히 조정할 수 있는 설계가 요구된다.

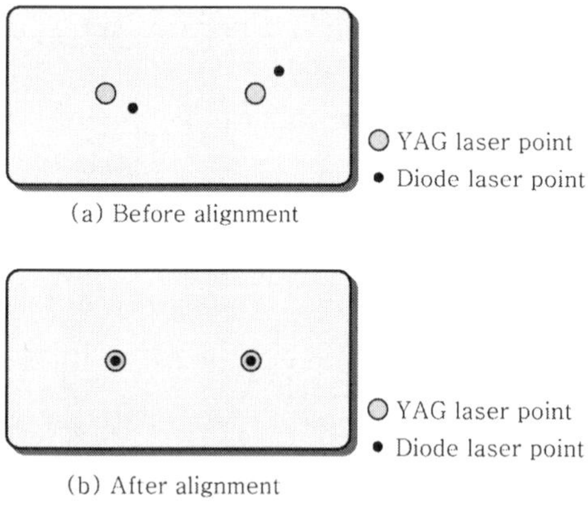

그림 2.B · 3 기능요구의 그림에 의한 설명

2.B.2 기존설계의 공리적 분석

독립공리를 이용하여 빔 정렬장치를 설계하기 위해서는 먼저 기능요구 벡터 FR과 설계파라미터 벡터 DP를 결정하고 이들간의 관계에 설계행렬 A를 사용한다. 특히 독립적인 기능요구의 정의는 설계분석에 있어서 상당히 중요한 비중을 차지한다. 이미 설계되어 사용되었던 빔 정렬장치의 기능요구는 다음과 같이 나타낼 수 있다.

FR_1 : 빔의 수직위치를 정렬시켜라.
FR_2 : 빔의 수직각도를 정렬시켜라.
FR_3 : 빔의 수평위치를 정렬시켜라.
FR_4 : 빔의 수평각도를 정렬시켜라.
FR_5 : 빔의 위치를 고정시켜라.

위와 같은 기능요구가 결정된 근거를 설명하면 다음과 같다. 먼저 빔의 경로를 서로 일치시킨다는 것은 빔의 두 중심축을 일치시킨다는 의미이다. 이러한 빔의 중심축을 일치시키려면 대상이 되는 빔 중심축의 연장선 위에 일치시키려는 빔의 시작점을 이동시켜야 한다(FR_1, FR_3). 시작점의 이동은 수직 및 수평위치가 모두 일치되어야 한다. 또한 빔의 거리에 관계없이 두 빔의 축이 일치하려면 두 빔의 각도가 서로 같아야 한다(FR_2, FR_4).

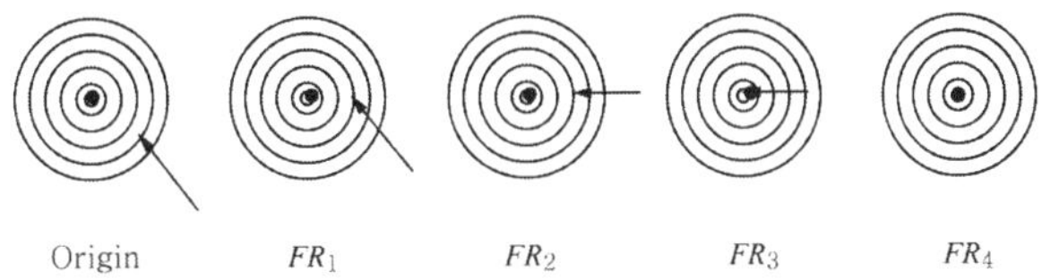

그림 2.B · 4 기능요구의 그림에 의한 설명

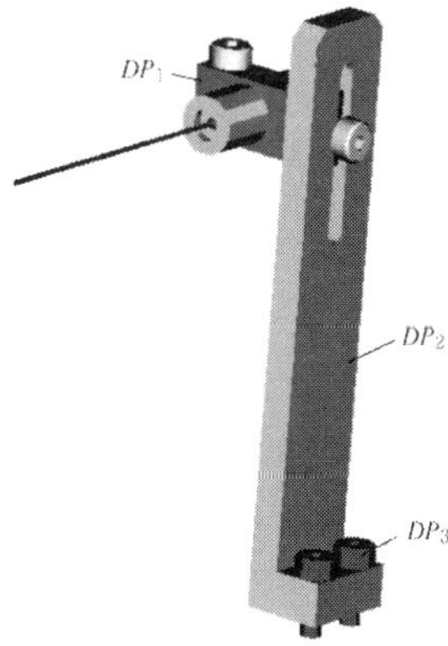

그림 2.B · 5 기존설계의 해

이러한 기능요구에 대한 정렬상태는 그림 **2.B · 4**와 같다. 기존의 설계가 그림 **2.B · 5**에 보여지고 있다. 기능요구에 대응되는 설계파라미터를 기존의 설계에서 찾으면 다음과 같다. 그림 **2.B · 5**에 각 설계파라미터를 표시하였다.

DP_1 : 수직이동블록

DP_2 : 지지블록

DP_3 : 하단나사

이에 대한 설계행렬은 식 (2.B · 1)과 같이 연성설계(coupled design)임을 알 수 있다.

$$\begin{bmatrix} FR_1 \\ FR_2 \\ FR_3 \\ FR_4 \\ FR_5 \end{bmatrix} = \begin{bmatrix} X & 0 & 0 \\ X & 0 & 0 \\ 0 & X & 0 \\ 0 & X & 0 \\ 0 & 0 & X \end{bmatrix} \begin{bmatrix} DP_1 \\ DP_2 \\ DP_3 \end{bmatrix} \qquad (2.B \cdot 1)$$

설계행렬을 구성해 본 결과 기존의 설계는 5개의 기능요구를 설계파라미터 3개로 조정해야 하는 연성설계임을 알 수 있다. 이 설계에서 수직위치를 정렬하고 싶을 때 수직이동블록을 움직여야 하므로 이미 설정된 수직각도가 같이 변동된다. 마찬가지로 수평위치를 정렬할 때 수평각도에도 영향을 미쳐 이를 동시에 정렬하기가 매우 어려운 구조로 되어 있다. 공리적설계에서 가장 이상적인 설계(ideal design)는 기능요구의 수와 설계파라미터의 수가 같아야 한다. 현재의 설계는 독립공리를 만족시키지 못하는 연성설계이므로 설계파라미터를 새로 추가하여 기능요구의 수와 같도록 하면 독립공리를 만족시키는 비연성설계 또는 비연성화설계를 도출할 수 있다.

2.B.3 공리적설계를 이용한 빔 정렬장치 개발

[1] 독립공리를 이용한 정렬장치의 설계안 도출

독립공리를 만족하는 새로운 설계를 도출하기 위하여 새로운 설계파라미터를 갖는 정렬장치를 고안하였다. 일단 수직위치 및 각도를 조절하는 FR_1, FR_2만을 고려하여 설계안을 도출한다면 수평위치를 조절하는 FR_3, FR_4는 같은 원리로 확장시킬 수 있다. FR_1, FR_2에 대한 기존 설계행렬을 다시 표시하면 식 (2.B · 2)와 같이 된다.

$$\begin{bmatrix} FR_1 \\ FR_2 \end{bmatrix} = \begin{bmatrix} X \\ X \end{bmatrix} [\, DP_1] \qquad (2.B \cdot 2)$$

위와 같은 연성설계를 개선하여 수직위치와 수직각도를 결정하기 위한 설계파라미터가 따로 구분되는 설계를 생각할 수 있다. 그 결과 다음과 같은 두 가지의 기본적인 설계안을 도출해 본다. 첫 번째 설계안은 그림 **2.B · 6**과 같이 빔이 시작하는 수직위치를 조절할 수 있는 뒤쪽 체결부와 수직각도를 조절할 수 있는 앞쪽 체결부로 구성된 설계파라미터를 가지고 있다. 이것은 각각의 구멍에 나사를 체결할 수 있고 조절 후 각 위치를 고정시킬 수 있는 설계다. 이에 대한 설계행렬을 구성한다면 다음과 같이 된다. 이는 이미 사용되고 있는 설계와 달리 식 (2.B · 3)과 같은 비연성화설계(decoupled design)가 된다.

$$\begin{bmatrix} FR_1 \\ FR_2 \end{bmatrix} = \begin{bmatrix} X & 0 \\ X & X \end{bmatrix} \begin{bmatrix} DP_1 \\ DP_2 \end{bmatrix} \qquad (2.B \cdot 3)$$

독립공리를 만족시키는 설계는 한 가지 이상이 나올 수 있다. 그러므로 독립공리를 만족시키는 다른 설계안을 고안한 결과 그림 **2.B · 7**과 같은 두 번째 설계안을 도출해 본다. 역시 빔의 수직 시작위치를 조정하는 뒤쪽 나사부와 빔의 수직각도를 조절하는 앞쪽 나사부가 분리되어 있다. 이 설계가 그림 **2.B · 6**과 다른 점은 빔의 각 위치를 직접 이동시키지 않고 나사를 이용하여 작은 양만큼 조절할 수 있다는 점이다. 이 설계안에는 위쪽 나사가 위치를 조정할 때 나사부 아래에는 스프링 플런저(spring plunger)가 부착되어 있다. 그러므로 직접 이동하지 않고 나사를 돌리는 것만으로 빔의 위치가 변동된다. 이 설계 또한 설계행렬을 표시하면 식 (2.B · 4)와 같은 비연성화설계가 된다.

$$\begin{bmatrix} FR_1 \\ FR_2 \end{bmatrix} = \begin{bmatrix} X & 0 \\ X & X \end{bmatrix} \begin{bmatrix} DP_1 \\ DP_2 \end{bmatrix} \qquad (2.B \cdot 4)$$

위의 두 가지 설계안은 각각의 기능요구에 대한 설계파라미터를 새로 정의하여 비연성화하는 설계이다. FR_1, FR_2를 비연성화시키는 개념을 같은 방법으로 확장시키면 다음과 같은 설계를 도출할 수 있다.

그림 **2.B · 8**은 그림 **2.B · 7**의 설계를 확장한 설계를 나타내고 있다. 이 설계의 기능요구와 설계파라미터를 정리하면 다음과 같이 된다.

FR_1 : 빔의 수직위치를 정렬시켜라.
FR_2 : 빔의 수직각도를 정렬시켜라.
FR_3 : 빔의 수평위치를 정렬시켜라.
FR_4 : 빔의 수평각도를 정렬시켜라.

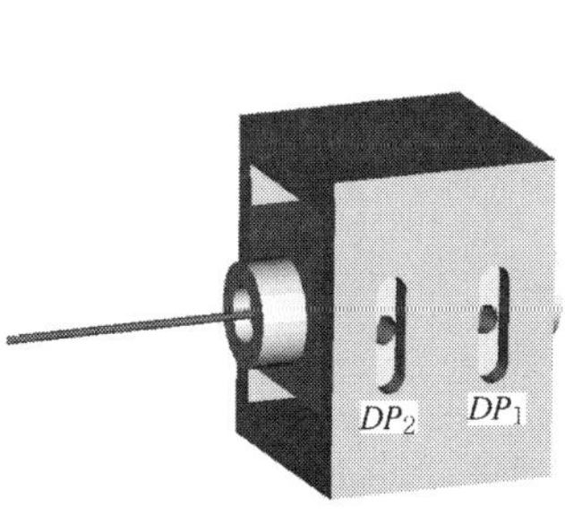

그림 2.B · 6 설계안 1

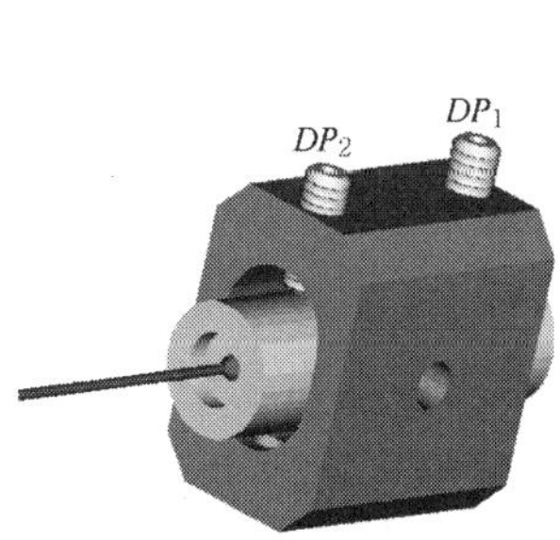

그림 2.B · 7 설계안 2

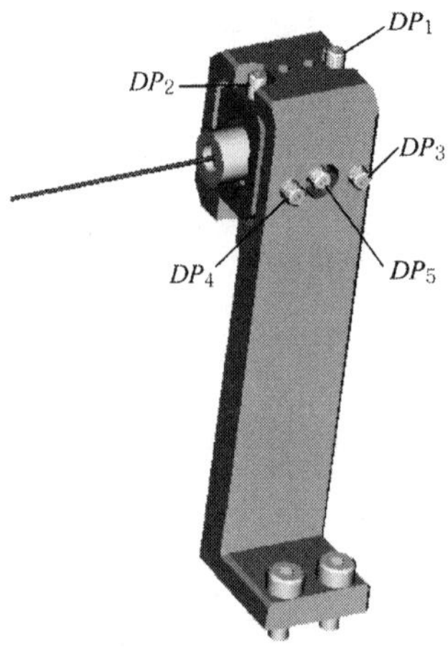

그림 2.B · 8 최종 설계안

FR_5 : 빔의 위치를 고정시켜라.

DP_1 : 위쪽 뒷나사

DP_2 : 위쪽 앞나사

DP_3 : 옆쪽 뒷나사

DP_4 : 옆쪽 앞나사

DP_5 : 고정 나사

여기서 DP_3과 DP_4는 앞에서 언급한 바와 같이 DP_1과 DP_2의 설계와 비슷한 것으로, 그림 **2.B · 6**과 **2.B · 7**에 보이는 두 가지의 설계를 생각할 수 있다. 그림 **2.B · 8**의 설계는 편의상 그림 **2.B · 7**의 방식으로 DP_3과 DP_4를 선정했다. 따라서 확장한 새로운 설계 역시 식 (2.B · 5)과 같은 비연성화설계행렬이 된다.

$$\begin{bmatrix} FR_1 \\ FR_2 \\ FR_3 \\ FR_4 \\ FR_5 \end{bmatrix} = \begin{bmatrix} X & 0 & 0 & 0 & 0 \\ X & X & 0 & 0 & 0 \\ 0 & 0 & X & 0 & 0 \\ 0 & 0 & X & X & 0 \\ 0 & 0 & 0 & 0 & X \end{bmatrix} \begin{bmatrix} DP_1 \\ DP_2 \\ DP_3 \\ DP_4 \\ DP_5 \end{bmatrix} \tag{2.B · 5}$$

따라서 도출된 설계는 기존의 설계보다 주어진 기능을 보다 독립적으로 만족시킨다고 할 수 있다. 물론 이 외에도 설계자에 따라 비연성화설계행렬을 가지는 다양한 설계 결과를 도출할 수 있다.

[2] 정보공리를 이용한 빔 정렬 장치의 선정

앞에서 도출한 두 가지의 설계 중 더 좋은 설계를 선택하기 위해 정보공리를 이용한다. 공리적 관점에서는 정보량이 적은 설계가 더 좋은 설계이다. 그림 **2.B · 6**과 **2.B · 7**에 있는 두 가지 설계의 정보량을 비교하기 위해 목표치 부근에서 각 설계파라미터의 분포에 대한 성공의 확률을 비교한다.

설계안 1은 조정 시에 이동거리가 빔의 이동거리와 일치하므로 그림 **2.B・9**에서 기울기(m)는 1이 된다. 반면에 설계안 2는 나사의 지름을 한바퀴 돌릴 때 다음 식과 같이 나사의 피치(pitch)만큼 빔이 이동한다.

$$2\pi r \tan\theta = p \qquad (2.B \cdot 6)$$

r : 나사의 반경

p : 나사산의 피치

θ : 나사산의 각도

반경 1.5mm에 피치가 1mm인 나사를 사용한 경우 기울기는 0.106이 된다. 빔 조정장치의 치수를 고려하여 두 가지 설계안에 대한 설계행렬을 구하면 다음과 같다.

$$\text{설계안 1} : \begin{bmatrix} FR_1 \\ FR_2 \end{bmatrix} = \begin{bmatrix} 1 & 0 \\ -4 & 4 \end{bmatrix} \begin{bmatrix} DP_1 \\ DP_2 \end{bmatrix} \qquad (2.B \cdot 7)$$

$$\text{설계안 2} : \begin{bmatrix} FR_1 \\ FR_2 \end{bmatrix} = \frac{1}{3\pi} \begin{bmatrix} 1 & 0 \\ -4 & 4 \end{bmatrix} \begin{bmatrix} DP_1 \\ DP_2 \end{bmatrix} \qquad (2.B \cdot 8)$$

정보량을 구하기 위해 목표값 $FR^* = [0 \quad 0]^T$는 두 설계에서 $DP^* = [0 \quad 0]^T$에 의해 만족되었다고 가정한다. 정보량은 식 (2・40)이나 식 (2・41)에 의해 도식적 방법으로 구할 수 있다. 이때 $\Delta FR_1 = 0.2\text{mm}$, $\Delta FR_2 = 0.3 rad$이라 가정하고 설계파라미터의 공차는 사용자의 손으로 조정할 때의 공차로서 $\Delta DP_i = 1\text{mm}$, $i = 1, 2$의 크기로 균일분포를 한다고 가정한다. 가정에 의해 성공의 확률의 크기가 달라질 수 있다. 여기서는 두 설계를 비교하는 것이므로 크기보다는 비교 결과가 중요하다. 따라서 위의 가정은 비교만을 위해서 유효하다고 볼 수 있다.

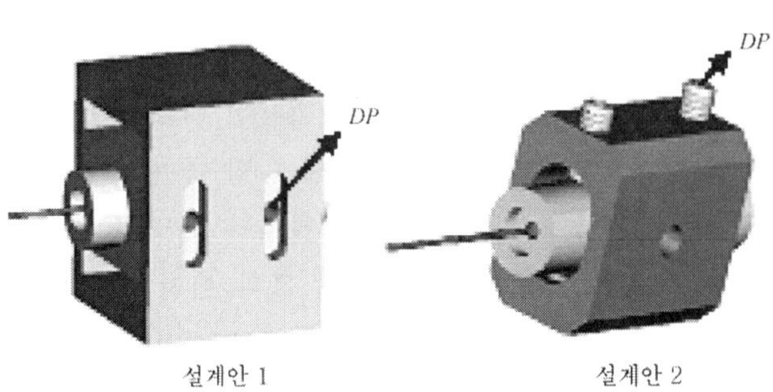

그림 2.B・9 정보량 비교를 위한 설계파라미터

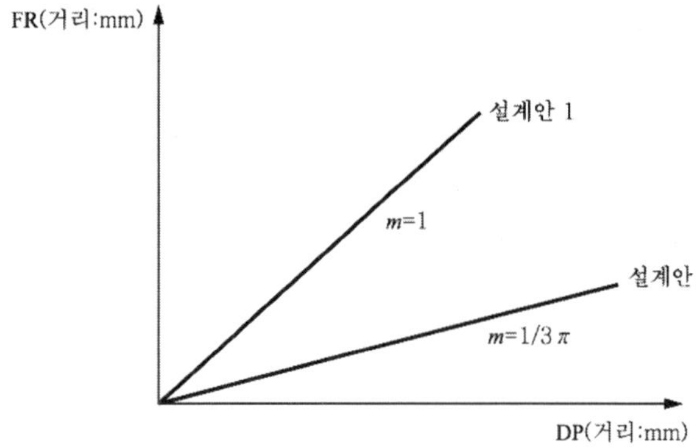

그림 2.B · 10 FR에 대한 DP의 경사도

도식화에 의한 방법은 그림 **2 · 13**과 그림 **2 · 14**의 방식으로 시행할 수 있다. 여기서는 적분법에 의한 성공의 확률을 계산해 본다. 그림 **2 · 15**에 있는 단위계단함수를 사용하여 각각의 설계안에 대한 확률분포를 식 (2.B · 9)과 같이 구할 수 있다.

$$p_i = \frac{1}{2\times1}[u(\delta DP_i-(-\Delta DP_i)) - u(\delta DP_i-(\Delta DP_i))],\quad i=1,2 \qquad (2.B \cdot 9)$$

식 (2.B · 7)~(2.B · 9)를 식 (2 · 49)에 대입하여 각각의 성공의 확률 p_{1s}, p_{2s}과 정보량 I_1, I_2를 구하면

설계안 1 : $p_{1s}=0.015$, $I_1=6.059$(bits)
설계안 2 : $p_{2s}=0.582$, $I_2=0.781$(bits)

이다. 따라서 설계안 2가 더 좋은 설계임을 알 수 있다. 이는 DP_3과 DP_4의 설계에서도 마찬가지이다. 결론적으로 설계안 2를 확장시킨 그림 **2.B · 8**의 설계를 최종설계로 확정한다.

2.B.4 요 약

여기서 보여진 예제는 그림 **2 · 20**의 흐름도가 적용된 대표적인 예이다. 즉, 독립공리에 의해 복수의 설계안을 도출한 후, 정보공리에 의해 가장 좋은 설계안을 선택하는 것이다. 그 결과로 기존의 빔 정렬장치의 단점을 해결한 새로운 개념의 우수한 설계가 도출되었다. 여기서의 설계대로 시제품을 제작하여 시험한 결과 이전의 설계에 비해 매우 만족스러운 성능을 보였다. 이상에서 본 바와 같이 기계장치의 개념설계에 공리적설계를 사용하면 좋은 설계를 얻을 수 있음을 알 수 있다(신광섭 · 박경진, 2004).

2.C TV 유리벌브 설계 시스템 개발

2.C.1 문제의 설명

TV나 컴퓨터에 사용되는 모니터의 출력장치로 사용되는 브라운관용 유리는 유리벌브(glass bulb)라 부른다(Do and Park, 2001). 그 재질이 유리로 구성되어 있고 제품에는 섀도 마스크(shadow mask), 전자총(electron gun)과 같은 장치가 부착되어 있다. 그리고 캐비닛(cabinet)에 고정시키기 위한 밴드(band)가 장착되어야 비로소 브라운관이라 부르게 된다. 결국 유리벌브 자체는 하나의 부품으로 취급되는 단위요소 제품으로 전면 유리(panel)와 후면 유리(funnel)의 두 가지로 구성된다. 그림 **2.C · 1**은 유리벌브의 형상을 표현하고 있다.

제품설계의 수행에서는 생산을 위한 정보 체계가 도면을 통해 이루어지고 있다. 그리고 각종 설계 과정을 분산된 여러 시스템에서 수행하는 등 비효율적으로 진행되고 있다. 이를 개선하는 소프트웨어를 제작함에 있어, 공리직실계를 이용하여 개발하려는 소프트웨어의 성능을 향상시키고 각종 설계단계에서의 정보 전달 체계를 강화하고자 한다.

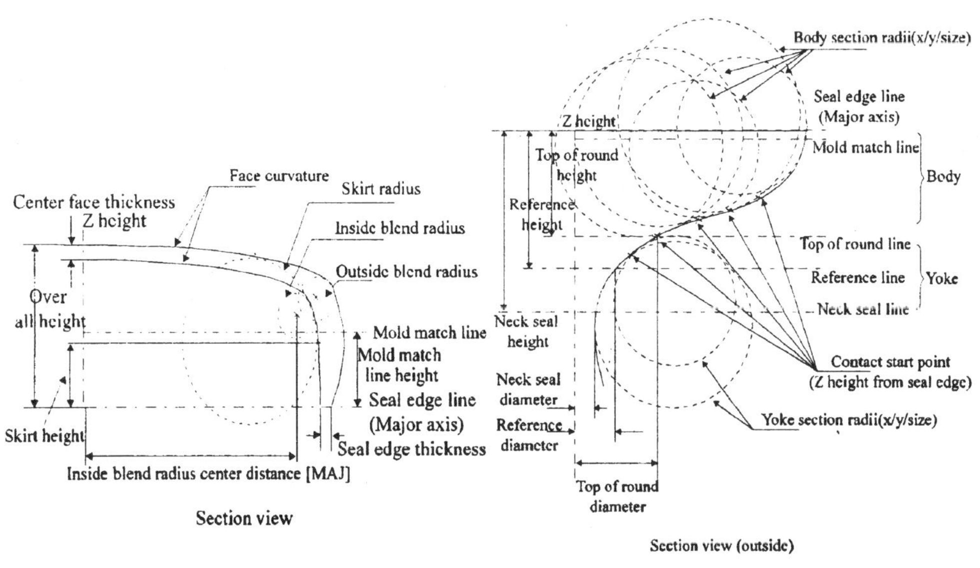

(a) Panel (b) Funnel

그림 2.C · 1 유리벌브의 형상

2.C.2 기존의 제품설계 과정의 공리적 분석

기존의 제품설계 과정은 그림 **2.C · 2**와 같다. 이 과정을 기능요구로 정의하면 아래와 같다.

FR_1 : 제품의 기초 정보를 구축하라.

FR_2 : 제품의 형상을 구축하라.

FR_3 : 제품의 성능을 검토하라.

FR_4 : 제품 도면을 생성하라.

이를 물리적 영역의 설계파라미터로 사상하면 다음과 같다.

DP_1 : 기초설계 데이터 집합

DP_2 : 제품의 3차원 기하형상 구조(panel/funnel)

DP_3 : 하중 및 외부조건(panel/funnel)

DP_4 : 도면 데이터 집합

위의 기능요구와 설계파라미터의 관계를 설계방정식으로 표현하면 식 (2.C · 1)과 같다.

$$\begin{bmatrix} FR_1 \\ FR_2 \\ FR_3 \\ FR_4 \end{bmatrix} = \begin{bmatrix} X & 0 & 0 & X \\ X & X & 0 & X \\ X & X & X & X \\ X & X & 0 & X \end{bmatrix} \begin{bmatrix} DP_1 \\ DP_2 \\ DP_3 \\ DP_4 \end{bmatrix} \qquad (2.C \cdot 1)$$

FR_1의 구현은 거래선에서 제공하는 기초 데이터와 제품 도면을 이용한다. 마찬가지로 FR_4의 구현도 기초 데이터, 제품의 3차원 기하형상구조, 제품 도면을 이용한다. 결국 FR_1과 FR_4는 현재의 설계방식을 따른다면 연성설계가 된다. 이를 개선하기 위하여 기능요구와 설계파라미터를 고찰해 보도록 하자.

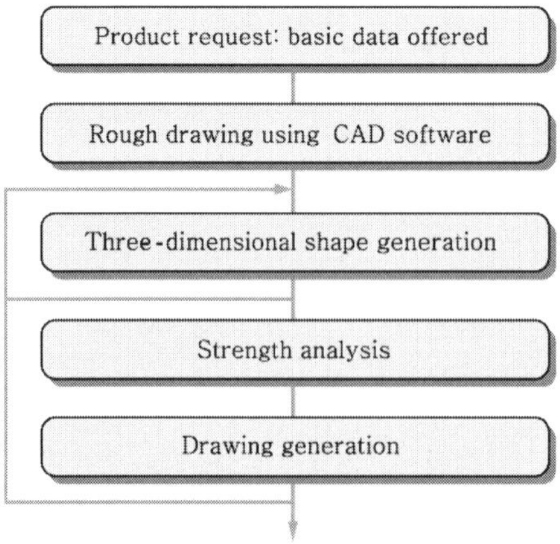

그림 2.C · 2 유리벌브 제품의 설계 과정

2.C.3 공리적설계에 의한 제품설계 자동화 소프트웨어의 구성

기능요구를 재정의하면 다음과 같다.

FR_1 : 신제품에 대한 데이터베이스를 생성하라.

FR_2 : 제품의 형상을 구축하라.

FR_3 : 제품의 성능을 검토하라.

FR_4 : 제품 도면을 생성하라.

이에 대한 설계파라미터를 선정하면

DP_1 : 신제품 데이터(panel/funnel)

DP_2 : 제품의 3차원 기하형상 구조(panel/funnel)

DP_3 : 하중 및 외부조건(panel/funnel)

DP_4 : 액세서리 데이터 집합

과 같다. 이에 대하여 설계행렬은 식 (2.C · 2)와 같이 비연성화설계로 생성된다.

$$\begin{bmatrix} FR_1 \\ FR_2 \\ FR_3 \\ FR_4 \end{bmatrix} = \begin{bmatrix} X & 0 & 0 & 0 \\ X & X & 0 & 0 \\ X & X & X & 0 \\ X & X & 0 & X \end{bmatrix} \begin{bmatrix} DP_1 \\ DP_2 \\ DP_3 \\ DP_4 \end{bmatrix} \tag{2.C · 2}$$

DP_1과 DP_2가 결정되면 FR_3은 상용 구조해석 소프트웨어를 이용하여 해결 가능하므로 FR_3은 하나의 모듈(M3)로 구성할 수 있다. 나머지 기능요구들은 비록 비연성화설계일지라도 설계파라미터의 정보가 명확하지 않으므로 주어진 설계파라미터를 만족하기 위한 새로운 기능요구를 제안하는 분해 과정이 필요하다.

FR_1에 대하여 하위수준의 기능요구를 선정하면

FR_{11} : 신제품의 ID 번호를 부여하라.

FR_{12} : 신제품의 데이터 집합을 구축하라.

이다. 마찬가지로 FR_2, FR_4에 대한 하위수준의 기능요구를 설정하면 다음과 같다.

FR_{21} : 곡률을 검토하라(panel 편평도, funnel 축별 profile).

FR_{22} : 3차원 형상을 계산하라.

FR_{23} : 제조 가능성을 검토하라.

FR_{41} : 제품의 형상을 표현하라.

FR_{42} : 주변 액세서리(accessory)를 표현하라.

두 번째 수준의 기능요구를 만족시키기 위한 설계파라미터를 선정하고 설계방정식을 세우면 다음과 같다.

DP_{11} : 신제품의 대표코드

DP_{12} : 신제품의 조건 데이터 집합

$$\begin{bmatrix} FR_{11} \\ FR_{12} \end{bmatrix} = \begin{bmatrix} X & 0 \\ X & X \end{bmatrix} \begin{bmatrix} DP_{11} \\ DP_{12} \end{bmatrix} \quad (2.C \cdot 3)$$

신제품의 ID 번호는 설계조건 데이터를 만들기 전에 이용하므로 식 (2.C・3)과 같이 비연성화설계가 도출된다. FR_{21}, FR_{22}, FR_{23}에 대한 설계파라미터는 다음과 같고 이에 대한 설계행렬이 식 (2.C・4)에 있다.

DP_{21} : 제품의 내외면 곡률

DP_{22} : 제품의 기하학적인 연관 관계식

DP_{23} : 금형제작에 관련된 데이터 집합

$$\begin{bmatrix} FR_{21} \\ FR_{22} \\ FR_{23} \end{bmatrix} = \begin{bmatrix} X & 0 & 0 \\ X & X & 0 \\ X & X & X \end{bmatrix} \begin{bmatrix} DP_{21} \\ DP_{22} \\ DP_{23} \end{bmatrix} \quad (2.C \cdot 4)$$

FR_{23}을 구현하는 방법 중 DP_{23}의 정의가 모호하므로 세분화작업이 필요하다. 따라서 FR_{23}에 대한 하위수준에 대한 정의가 필요하다. FR_{41}, FR_{42}에 대한 설계파라미터는 다음과 같다.

DP_{41} : 제품설계 데이터 집합

DP_{42} : 액세서리 데이터 집합

이에 대한 설계행렬은 다음과 같다.

$$\begin{bmatrix} FR_{41} \\ FR_{42} \end{bmatrix} = \begin{bmatrix} X & 0 \\ 0 & X \end{bmatrix} \begin{bmatrix} DP_{41} \\ DP_{42} \end{bmatrix} \quad (2.C \cdot 5)$$

FR_{23}에 대하여 하위수준의 기능요구를 정의하면 다음과 같다.

FR_{231} : 전면 유리의 유효 화면거리(useful screen dimension)를 만족하는가 검토하라.

FR_{232} : 방출성(ejectability)이 용이하게 하라.

FR_{233} : 후면 유리의 주사선 편향각도가 만족되는지 검토하라.

이를 만족하는 설계파라미터는 다음과 같다.

DP_{231} : 내면 블렌딩 원(blending circle) 중심점의 중심축으로부터의 거리

DP_{232} : 측면 벽(side wall) 각도

DP_{233} : 요크부 내면 곡률

설계행렬은 다음과 같이 정의된다.

$$\begin{bmatrix} FR_{231} \\ FR_{232} \\ FR_{233} \end{bmatrix} = \begin{bmatrix} X & 0 & 0 \\ 0 & X & 0 \\ 0 & 0 & X \end{bmatrix} \begin{bmatrix} DP_{231} \\ DP_{232} \\ DP_{233} \end{bmatrix} \qquad (2C \cdot 6)$$

FR_{23}의 하위수준은 비연성설계이며, 결과적으로 FR_2를 구성하는 하위수준의 설계는 비연성화설계이다.

그림 **2.C · 3**은 유리벌브 제품설계 자동화 시스템에 대한 모듈 구성을 나타내는 모듈결합구조도이다. 또한 그림 **2.C · 4**는 설계의 흐름을 알 수 있는 정보흐름도이다. 그림 **2.C · 4**에서 점선으로 표시된 선은 기능요구와 설계파라미터들 간의 분해 과정에 대한 반복결합을 의미하는 것이 아니다. 다만 설계의 흐름상 해석 결과가 좋지 않아 설계파라미터의 값을 변경하는 반복을 나타낸다.

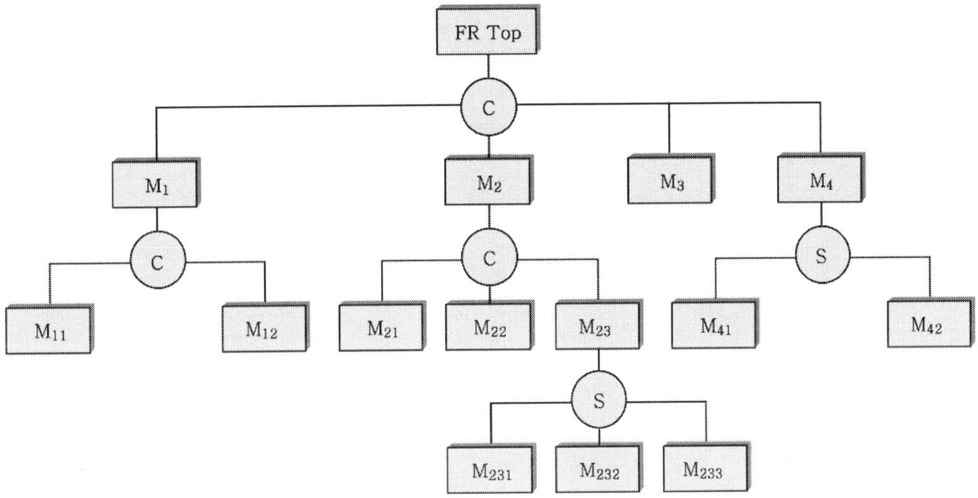

그림 2.C · 3 유리벌브 제품설계 자동화 시스템의 모듈결합구조도

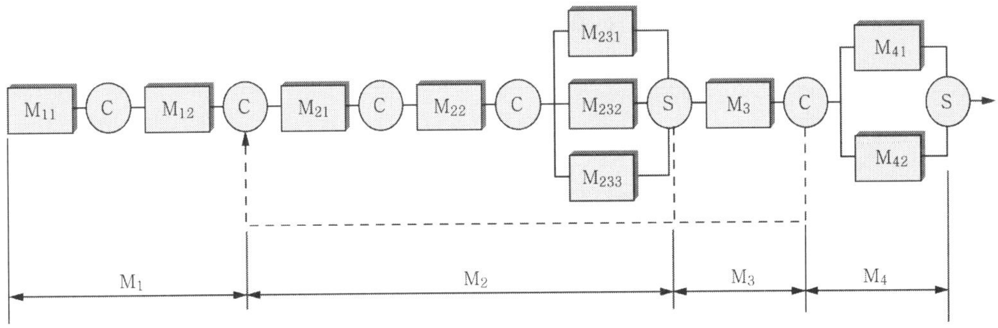

그림 2.C · 4 그림 2C · 3의 모듈결합구조도에서 유추된 정보흐름도

2.C.4 제품설계 자동화 소프트웨어의 개발

제품설계 자동화를 위한 소프트웨어를 실제로 구현하기 위해서 몇 가지 성격이 다른 부분들을 통합해야 한다. 따라서 실행파일의 크기가 커지는 단점이 있다. 모듈별로 프로그램을 공용 라이브러리(shared library) 형태로 만드는 방식을 추천한다. 설계를 위해 많은 곡률 데이터 및 위치 데이터 등을 운영하기 때문에 처리하는 데이터의 양이 매우 많아 메모리 공간을 확보하기 어렵다. 이에는 동적 메모리 할당 기법을 사용하여 필요 시에 메모리를 할당해서 쓴다. 소프트웨어의 기능을 간략하게 설명한다.

[1] 사용자 인터페이스

사용자 인터페이스는 UNIX 시스템의 표준 그래픽 도구인 X-window(MOTIF)를 이용하여 소프트웨어를 개발한다(Heller, 1994). 사용자 인터페이스를 통해 각종 메뉴시스템이나 그래프 등의 표현 양식을 제공한다. 그림 **2.C · 5**는 사용자 인터페이스의 화면의 예를 보이고 있다.

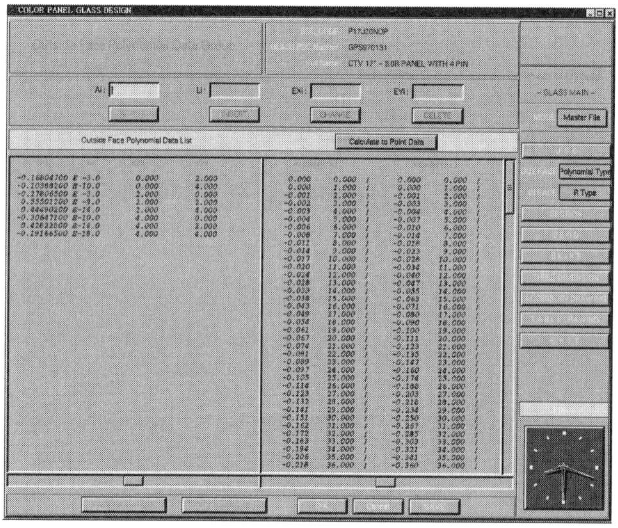

그림 2.C · 5 유리벌브 제품설계 자동화시스템의 메뉴 화면

[2] 데이터베이스

전체 제품설계 시스템의 지원을 위하여 데이터베이스(ORACLE) 접속 프로그램은 두 가지 방식을 사용한다(Oracle Co., 1990). 첫째는 사용자 인터페이스를 통해 정상적으로 데이터베이스를 접속하는 방법이고, 둘째는 비상시의 경우(예를 들어, 데이터베이스의 복

사 및 이전, 재구축 등)에 그래픽 사용자 인터페이스와는 상관없이 데이터베이스를 접속하도록 하는 방법이다. 접속 시 수행되는 기본 명령들은 메모리의 효율적인 이용을 위하여 접속방식에 관계없이 공용 라이브러리화하여 사용한다.

[3] 3차원 형상

전면 유리와 후면 유리의 3차원 형상 및 강도해석을 위한 메시를 생성하기 위해 자체 개발한 유리벌브 형상 계산 전용 프로그램인 BULB-3D를 이용한다(박경진 외, 1995). BULB-3D는 몇 가지 기하학과 수치 해법을 이용하여 유리벌브의 3차원 곡면형상 및 강도해석을 위한 메시를 생성한다. 3차원 형상은 그래픽 라이브러리인 스타베이스(Starbase)를 이용하여 축소/확대, 이동, 회전, 렌더링(rendering) 기능을 제공한다(Hewlett-Packard Co., 1991). 그림 **2.C·6**에 제품의 3차원 형상의 출력 예가 있다.

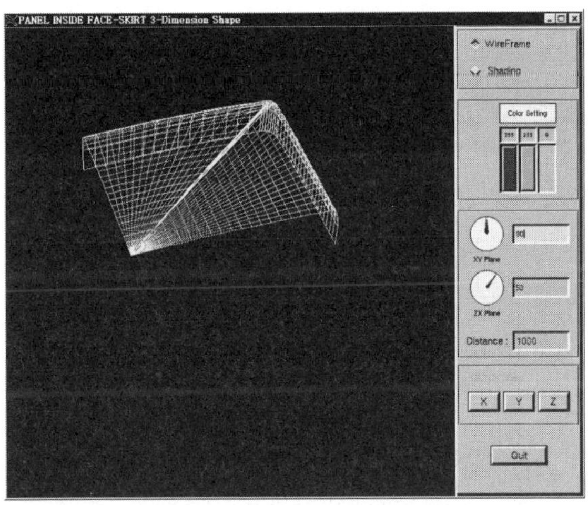

그림 2.C·6 3차원 형상에 대한 메시 생성 후의 출력 화면

[4] 강도해석

강도해석을 위해서 상용 해석 소프트웨어인 ANSYS를 이용하여 제품의 진공강도 및 밴드강도 성능을 검증한다(ANSYS Inc., 1993). 메시가 정의되면 유리의 물성치, 하중조건, 구속조건 등의 데이터를 바탕으로 제품의 강도를 평가한다. 원하는 강도 결과를 얻을 때까지 3차원 형상 생성모듈과 연계해서 각종 변수를 바꿔 가면서 반복적으로 수행한다. 그림 **2.C·7**에 강도해석의 결과 화면이 있다.

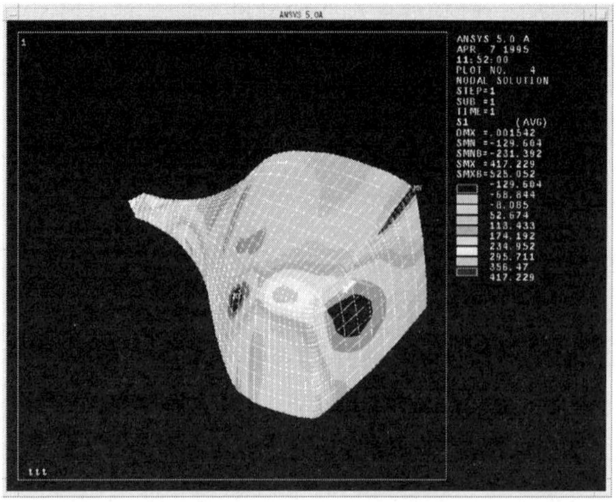

그림 2.C · 7 유리벌브의 강도해석 결과의 출력 화면

[5] 도면출도

도면출도를 위해서는 유니그래픽스(Unigraphics)라는 상용 CAD 소프트웨어를 이용한다(Electronic Data Systems Co., 1993). 유니그래픽스가 제공하는 프로그래밍 연결도구인 그립(GRIP)과 사용자 함수(user function)를 이용하여, 제품의 형상, 치수, 측정 위치, 핀 장착 위치, 정렬 위치, 표 등을 삽입하였다. 또한 사용자의 선택에 따라 도면의 항목을 조정할 수 있다. 자동 생성된 도면의 예가 그림 **2.C · 8**에 있다.

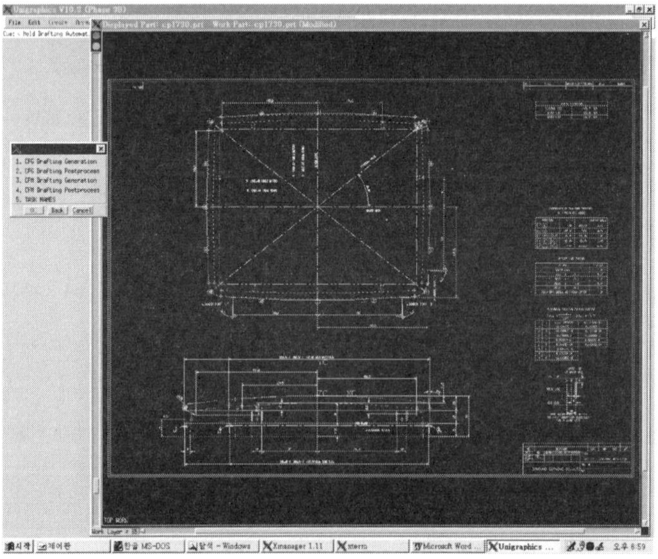

그림 2.C · 8 유리벌브의 도면출도 화면

2.C.5 요 약

공리적설계를 제품설계용 소프트웨어 설계에 어떻게 적용하는가에 대해 설명했다. 개발된 소프트웨어를 통하여 그간 설계자의 경험에만 의존하여 제품을 설계해 왔던 관행을 고쳤다. 또한 개념설계단계에서 상세설계단계인 도면작성까지 자동화하여 신제품의 개발기간을 단축했다. 공리적설계에 의하여 설계 과정을 분석하여 기능요구와 설계파라미터를 정의하고 분해함으로써 자동적으로 소프트웨어의 모듈이 정의되었다.

2.D 모니터용 EPS 완충 포장재 설계 시스템 개발

2.D.1 문제의 설명

모니터 제품은 운반 중에 취급 부주의로 인해서 낙하 및 충격 등의 많은 파손 위험에 노출된다(Yi and Park, 2005). 노출된 위험으로부터 제품을 안전하게 보호하기 위해 완충 포장을 하는데, 완충재로는 발포성 폴리스티렌(Expanded PolyStyrene, EPS)이 널리 쓰이고 있다. 그러나 완충 포장재의 제비용 부담뿐만 아니라 모니터 제품의 부피 증가로 인하여 물류 및 수송비용이 증가한다. 따라서 포장의 부피를 줄이면서 낙하 및 충격에 견딜 수 있는 적절한 완충 포장재 설계가 이루어져야 한다.

현장에서 완충재의 개념설계는 전적으로 설계자의 경험에 의존하고 있다. 시작제품을 제작한 후 낙하 충격 실험으로 검증하는 반복적인 시행착오 과정을 거친다. 이는 선임자의 잘못된 전례를 그대로 따를 수 있고 실험에 필요한 비용부담도 증가한다. 이러한 반복과정을 줄이기 위하여 충격해석 도구를 도입하고 있다. 그러나 사용상 기술 습득의 번거로움과 효율적인 설계를 위하여 개념설계단계에서 모니터 모델에 대한 완충 포장재 설계를 자동화하는 시스템이 필요하다.

이에 EPS 완충재의 유한요소해석모델을 상용소프트웨어인 LS/INGRID를 이용하여 자동으로 구성하고 LS/DYNA3D로 낙하해석을 실행하여 결과를 보여줌으로써 완충재를 보다 효과적으로 설계할 수 있는 설계 시스템을 개발하고자 한다(Livermore Software Technology Co., 1998 · 1999). 설계 시스템의 설계 목적을 공리적설계의 독립공리에 따라 정의하면 설계 과정이 순차적으로 진행될 수 있다. 따라서 자동설계 시스템 개발이 용이하다.

2.D.2 공리적설계를 이용한 모니터용 EPS 완충 포장 설계 시스템 설계 과정

2.5절에서 소개한 V-모델의 설계 과정을 완충 포장 설계 시스템을 설계하는 데 적용한다. 개발할 시스템은 모니터 제품의 포장재에 대한 개념설계단계에서 상세설계 수준의 형상치수까지 결정하는 자동설계 시스템이다. 우선 해당업체의 설계자로부터 지금까지 관행으로 해오고 있는 설계 과정을 습득하였다. 또한 지금까지 설계를 행하면서 필요하다고 생각하는 요구사항을 듣고 사용자요구(CAs)를 정리하였다. 시스템 개발 과정을 설명한다.

[1] 설계 시스템의 기능요구의 정의 및 분해(단계 1, 2, 3)

공리적 관점에서 EPS 완충 포장 설계의 목적을 정의한다. 최상위수준에서의 기능요구와 설계파라미터의 관계는 표 **2.D · 1**과 같다. 설계행렬의 형태가 삼각행렬이므로 비연성화설계이다. 그러므로 소프트웨어의 설계 과정은 설계행렬에 표현된 순서에 따라 이루어져야 한다. 각 기능요구는 선정된 설계파라미터를 기준으로 분해할 수 있다. 소프트웨어 관점에서 기능요구와 설계파라미터의 분해 과정은 최소단위의 알고리즘이나 메소드(method)가 도출될 때까지 진행한다.

FR_2의 기능요구를 분해하는 예를 보인다. 표 **2.D · 1**에 보이는 바와 같이 FR_2는 '모델링과 모사실험을 위한 데이터를 설정하라'이다. 그리고 FR_2에 대한 설계파라미터 DP_2는 모니터를 모델링하기 위한 입력 데이터와 해석을 위한 입력 데이터로 선정하였다. DP_2로부터 FR_2의 설계 과정을 세분화할 수 있다. 즉, 모니터를 모델링하기 위한 데이터, 해석을 위한 데이터, 물성 데이터 등의 다양한 입력 데이터를 사용자가 설정하는 기능을 정의해야 한다. 그러므로 FR_2는 하위수준에서 표 **2.D · 2**와 같이 분해할 수 있다. 분해 과정은 최하위수준까지 진행한다. 표 **2.D · 3**에 FR_2를 최하위수준까지 분해했을 때의 설계 관계가 있다. 소프트웨어 시스템의 설계 흐름은 소프트웨어의 옵션설정 및 데이터 관리기능을 제외하고는 실제 설계 과정과 같다.

표 2.D · 1 모니터용 EPS 완충 포장재 설계용 시스템에 대한 최상위 수준에서의 기능요구

	FR_x	DM	DP_x
1	옵션을 설정하라.	X O O O O	옵션 정보 데이터
2	모델링과 모사실험을 위한 데이터를 설정하라.	X X O O O	모델링 및 낙하 충격해석을 위한 데이터
3	완충재의 유한요소모델을 생성하라.	X X X O O	완충재의 설계변수
4	모사해석 후, 분석을 통하여 개선된 설계안을 제시하라.	X X X X O	DYNA3D의 입력 구문
5	설계 데이터를 관리하라.	X X X X X	데이터 관리자

표 2.D · 2 FR_2의 기능요구 분해

	$FR_{2.x}$		$DP_{2.x}$
	모델링과 모사실험을 위한 데이터를 설정하라.	DM	모델링 및 낙하 충격해석을 위한 데이터
1	모니터의 모델링 데이터를 설정하라.	X O O	모니터의 모델링 데이터
2	완충재의 모델링 데이터를 설정하라.	O X O	완충재의 모델링 데이터
3	낙하 충격해석을 위한 조건을 설정하라.	O O X	낙하 조건

표 2.D · 3 최하위수준에서 FR_{21223}의 기능요구 분해

	$FR_{21223.x}$		$DP_{21223.x}$
	노드와 요소 정보 파일을 DYNA3D 형식의 파일로 변환하여라.	DM_{21223}	노드와 요소 설정에 관한 DYNA3D 형식 정보
1	노드와 요소 정보 파일을 읽어라.	X O O O	파일 이름
2	노드와 요소 번호를 충분히 크게 재설정하라.	X X O O	노드와 요소 번호
3	DYNA3D 형식으로 변환하라.	X X X O	DYNA3D 형식 구문
4	노드의 재설정 번호를 저장하라.	O X O X	노드의 재설정 번호

[2] 모듈 정의 및 객체 선정(단계 4, 5)

분해의 지그재그 과정을 통하여 전체 설계행렬을 구성할 수 있다. 전체 설계행렬은 소프트웨어의 모듈 및 객체를 정의하는 데 중요한 역할을 한다. 그림 **2.D · 1**의 FR_3에 대한 설계행렬을 살펴보도록 하자. 그림에서와 같이 설계자는 설계행렬을 통하여 굵은 선으로 표시한 사각행렬의 객체를 정의한다. 객체가 수행해야 할 기능인 기능요구는 소프트웨어의 모듈을 정의할 수 있다. 또한 각 기능요구에 해당하는 설계파라미터는 객체가 가진 속성(attributes)으로 입력 및 출력 데이터에 해당한다. 객체는 물리적 영역의 속성을 조절하여 기능적 영역의 기능을 수행한다. 이를 객체 지향적 소프트웨어에서는 행위(behavior)로 정의하고, 각 행위는 소프트웨어 코딩(coding) 시 메소드(methods)로 정의한다. 그러므로 설계자는 설계행렬을 통하여 기능적 영역과 물리적 영역 사이의 관계를 알 수 있다.

개발한 소프트웨어는 주모듈, 데이터 관리모듈, 모델링모듈, 낙하 충격해석 모듈, 결과분석 및 설계모듈로 구성되어 있다. 주모듈의 기능은 입력 및 전체설계 과정을 사용자가 제어하는 것이다. 데이터 관리 모듈은 시스템의 데이터를 관리하는 역할이다. 이는 외부 실행, 물성 데이터를 담고 있는 파일, 표준 직교배열표에 대한 데이터베이스를 관리한다. 모델링 모듈은 해석을 위한 입력 데이터를 파일로 생성하는 기능으로 완충재의 형상 크기,

상용 도구인 LS/INGRID를 통해 자동으로 유한요소해석모델을 생성하여 해석을 위한 입력 파일을 작성한다. 낙하 충격해석 모듈은 상용 해석기인 LS/DYNA3D의 입력 파일을 자동으로 생성하고 낙하 충격해석을 실행하도록 하는 기능을 수행한다. 결과분석 및 설계 모듈은 낙하 충격해석을 실행한 결과 데이터를 분석하고 새로운 설계안을 제시한다.

객체는 속성과 메소드로 정의한다. 설계자는 설계행렬의 사각행렬을 구성한다. 세로열에 있는 설계파라미터를 속성으로 하고 가로행에 표현한 기능요구를 메소드로 하는 하나의 객체를 결정할 수 있다. 또한 여러 개의 객체들을 추상화(abstraction)한 새로운 객체를 정의할 수 있다. 예를 들어, 그림 **2.D·1**에서 완충 포장재의 유한요소해석모델을 생성하는 기능을 수행하는 FR_3 모듈은 객체 A로 정의할 수 있다. 그런데 객체 A는 완충재의 내부 형성을 정의하는 객체 B와 완충재의 외부형상을 정의하는 객체 C를 포함하고 있다. 모니터와 완충재가 서로 접촉하는 부분은 네 곳이 있다. 객체 B는 5개의 객체를 포함하고 있는데 완충재의 접촉면에 대한 형상을 결정하는 것이다. 네 곳의 접촉면에 대한 완충재의 형상을 결정하는 4개의 객체들과 공통적으로 사용할 수 있는 입력 데이터를 설정하는 1개의 객체를 포함하고 있다. 같은 방법으로 객체 C는 리브의 형상에 따라 5개의 객체로 구성하였다. 이와 같이 정의한 객체는 다음 단계에서 클래스로 정의할 수 있다.

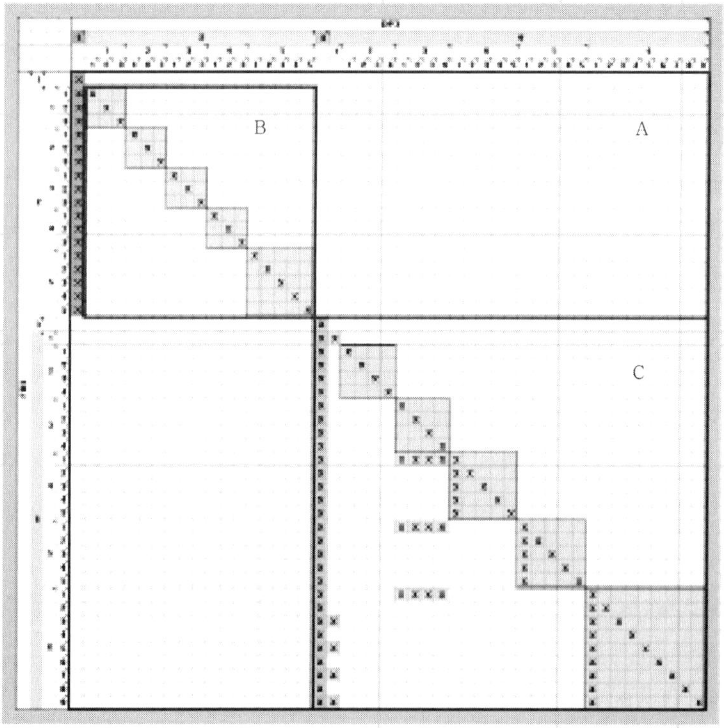

그림 2.D·1 FR_3에 대한 설계행렬

[3] 클래스 관계 정의 및 코딩(단계 6, 7)

그림 **2.D · 2**에 각 객체들에 대한 클래스를 정의하였다. 여기서는 하위수준의 클래스들은 표현하지 않았다. 클래스 'PackDesign'은 4개의 클래스들을 포함하고 있다. 클래스 'Option'은 초기에 사용자가 정의해야 하는 시스템의 환경 데이터를 설정하는 기능을 수행한다. 이는 시스템의 구동 시 자동으로 설정한다. 클래스 'BasicModel'은 모니터와 완충재의 물성치를 설정하며, 클래스 'CushionModel'은 완충재의 형상을 설정한다. 사용자에게서 입력받은 데이터들은 시스템에 의하여 좌표를 변환하여 자동으로 유한요소해석을 위한 입력 파일을 생성한다. 클래스 'IterationAnal'은 선정한 직교배열표에 따라 낙하 충격해석을 자동으로 실시하고 그 결과를 분석하여 사용자에게 알맞은 완충 포장재의 형상을 제안한다.

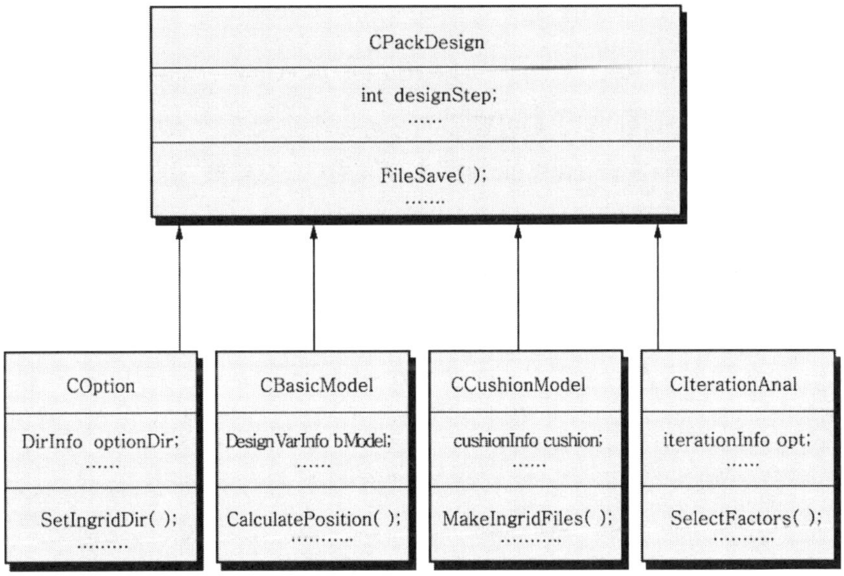

그림 2.D · 2 'PackDesign' 소프트웨어의 클래스 다이어그램

다음으로 정의한 클래스에 따라 프로그램을 작성하였다. 설계자는 클래스의 속성과 메소드에 해당하는 데이터와 알고리즘을 코딩한다. 그림 **2.D · 3**에서는 앞에서 설명한 과정을 통하여 개발한 소프트웨어로 설계하는 화면을 보이고 있다.

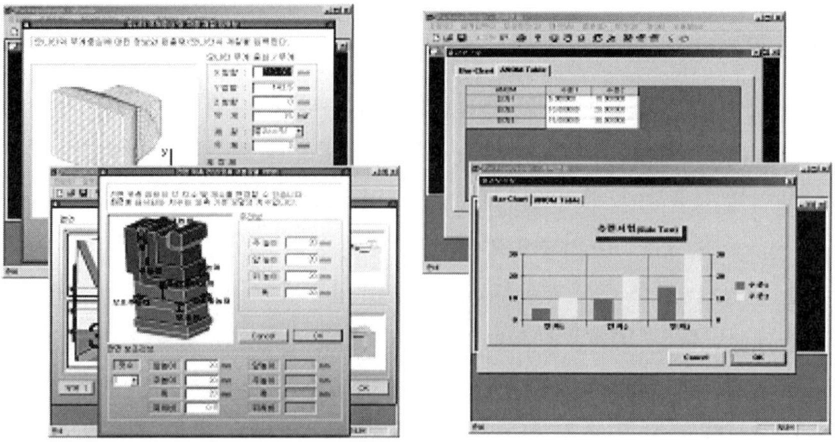

그림 2.D · 3 PackDesign 소프트웨어를 이용하여 설계한 결과 화면

2.D.3 요 약

공리적설계를 소프트웨어 설계에 어떻게 적용하는가에 대해서 설명했다. 소프트웨어 설계 방법인 객체 지향적 프로그래밍의 개념과 연계한 V-모델의 과정에 따라 개발하였다. 그리하여 소프트웨어 효율을 향상시킬 수 있었다. 여기서는 FR_1과 FR_2를 분해한 것만 보였으며 FR_3에 대해서는 결과행렬만 수록했다. FR_4와 FR_5는 설계 과정으로서 그에 대한 설명은 생략하였으나 분해 과정들은 다른 것과 비슷하다. 개발된 소프트웨어를 사용하면 설계경험이 부족한 설계자도 주어진 입력변수를 설정하여 새로운 모니터에 맞는 완충재를 쉽게 설계할 수 있다. 그리고 설계 결과를 저장하여 새로운 제품에 대한 설계에 반영할 수 있다.

찾 아 보 기